THE CONCH BOOK

by

DEE CARSTARPHEN

PUBLISHED BY PEN & INK PRESS
P.O. BOX 235, WICOMICO CHURCH, VA 22579
TEL/FAX: (804) 580-8723
www.peninkpress.com

THANKS TO

- ALL THE BAHAMIANS, WEST INDIANS, YACHTSMEN, AND OTHERS WHOSE INTEREST AND CURIOSITY ABOUT THE CONCH PROMPTED THIS WORK.

- DR. SCOTT SIDDALL, WHO, WITH DR. EDWIN IVERSEN, WORKED AT THE UNIVERSITY OF MIAMI'S SCHOOL OF MARINE AND ATMOSPHERIC SCIENCES ON *STROMBUS GIGAS* AQUACULTURE. DR. SIDALL WAS MOST HELPFUL IN DEMONSTRATING CONCH-REARING IN THE LABORATORY.

- DR. JOHN F. STORR, A BIOLOGIST AT THE STATE UNIVERSITY OF NEW YORK, BUFFALO, WHOSE THOUSANDS OF UNDERWATER HOURS MADE HIS REVIEW OF THE MANUSCRIPT ESPECIALLY WORTHWHILE.

- STUART K. HOPKINS, ABOARD WHOSE KETCH *SEA WIND* MOST OF THE RESEARCH FOR THIS BOOK WAS DONE. *SEA WIND* VISITED SUCH CENTERS OF CONCH INFORMATION AS MIAMI, KEY WEST, AND NASSAU, FREQUENTLY ANCHORING OVER CONCH BEDS TO FACILITATE FIRST-HAND OBSERVATIONS.

- NEIL SEALEY AND DR. KATHLEEN SULLIVAN SEALEY FOR THEIR SPECIAL HELP IN UPDATING THIS REVISED EDITION. MR. SEALEY IS DIRECTOR OF MEDIA PUBLISHING LTD, NASSAU. PROFESSOR SEALEY DIRECTS THE MARINE RESEARCH CENTER AT THE UNIVERSITY OF MIAMI.

- BLACK JACK, THE SHIP'S CAT, WHO ONE DAY FOUND A BEAUTIFUL BIG PINK PEARL IN HIS DINNER OF CONCH TRIMMINGS.

The QUEEN CONCH

In the blue-green shallows, the great conch moves slowly, humping herself along by extending her foot and digging her pointed sickle-shaped operculum into the sand. She works, out of the turtle grass (*THALASSIA*) that is her feeding place, where she dines on algae, and onto a patch of clear sand.

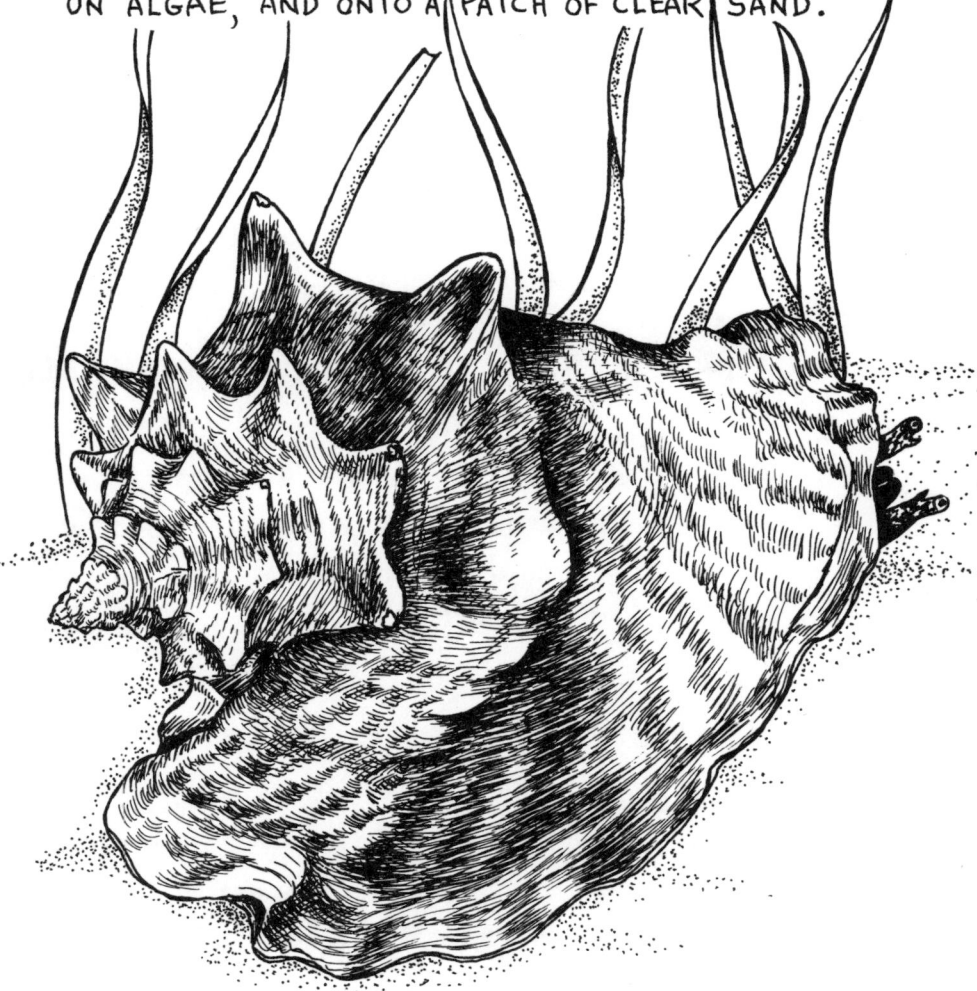

HERE SHE BEGINS TO SPIN FROM HER VAGINAL
FOLD A THIN, STICKY STRAND THAT RESEMBLES
MONOFILAMENT FISHING LINE IN WHICH A
THREAD OF EGGS IS COILED. SHE MANIPU-
LATES IT WITH HER FOOT TO COAT IT
LIGHTLY WITH SAND BEFORE FOLDING IT
BACK UPON ITSELF, BACK AND FORTH,
UP AND DOWN. SHE WORKS FOR
TWENTY-FOUR TO THIRTY-SIX HOURS,
BUILDING A SLIGHTLY CRESCENT-
SHAPED MASS, ABOUT THE SIZE OF A
SMALL FAT BANANA, WHICH RESTS IN A
DEPRESSION SHE'S MADE IN THE SAND. DURING
THIS TIME *STROMBUS GIGAS* NEITHER EATS NOR
RESTS. SHE DOES NOT GUARD HER EGGS AND AS
SOON AS SHE IS FINISHED, SHE "WALKS" AWAY.
THE MATING SEASON STARTED IN EARLY SPRING AND
WILL LAST UNTIL FALL. SHE MAY MATE MANY TIMES
THIS SEASON AND LAY SEVERAL EGG MASSES.

IF THIS GELATINOUS STRAND OF EGGS WERE
STRETCHED OUT, IT WOULD BE ABOUT ONE HUNDRED
FEET LONG. IT CONTAINS CLOSE TO A HALF MILLION
EGGS! AT THAT RATE, YOU MIGHT THINK CONCH
WOULD TAKE OVER THE WORLD, BUT THE SURVIVAL
RATE OF THE YOUNG IS SLIGHT.

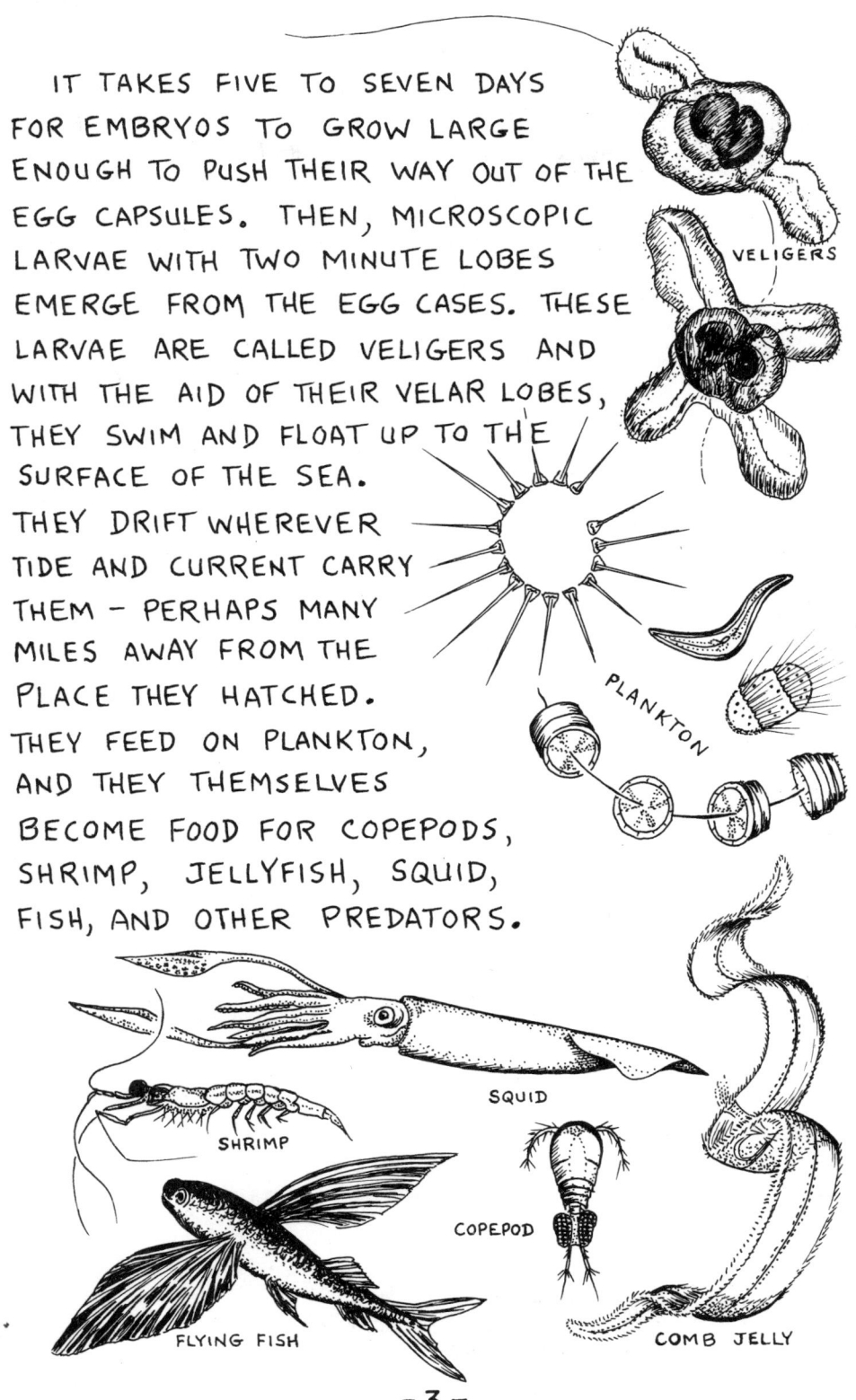

IT TAKES FIVE TO SEVEN DAYS FOR EMBRYOS TO GROW LARGE ENOUGH TO PUSH THEIR WAY OUT OF THE EGG CAPSULES. THEN, MICROSCOPIC LARVAE WITH TWO MINUTE LOBES EMERGE FROM THE EGG CASES. THESE LARVAE ARE CALLED VELIGERS AND WITH THE AID OF THEIR VELAR LOBES, THEY SWIM AND FLOAT UP TO THE SURFACE OF THE SEA. THEY DRIFT WHEREVER TIDE AND CURRENT CARRY THEM — PERHAPS MANY MILES AWAY FROM THE PLACE THEY HATCHED. THEY FEED ON PLANKTON, AND THEY THEMSELVES BECOME FOOD FOR COPEPODS, SHRIMP, JELLYFISH, SQUID, FISH, AND OTHER PREDATORS.

VELIGERS

PLANKTON

SQUID

SHRIMP

COPEPOD

FLYING FISH

COMB JELLY

-3-

THEY GROW AND CHANGE RAPIDLY.
EVEN AT THIS EARLY STAGE THEY
HAVE A TINY TRANSPARENT SHELL
CALLED A PROTOCONCH.
IN ABOUT A MONTH'S
TIME, WHEN THEY
ARE THE SIZE OF A
LARGE GRAIN OF SAND,
THEY METAMORPHOSE
FROM A PELAGIC
(FREE SWIMMING) TO A
BENTHIC (BOTTOM DWELLING) ORGANISM. THEY
SETTLE TO THE BOTTOM AND ABSORB THEIR
SWIMMING LOBES. THEY GROW A FOOT FOR
LOCOMOTION AND A MOUTH (PROBOSCIS) FOR
GRAZING.

AFTER
D'ASARO

BEING YOUNG AND TENDER, THEY BURY
THEMSELVES AROUND THE ROOTS OF THE
THALASSIA DURING THE DAY FOR PROTECTION.
THAT'S WHY VERY SMALL CONCH ARE RARELY
SEEN. THEY'VE BEEN LEADING A SECRET LIFE.
THEY ONLY LEAVE THEIR HIDING PLACE AT NIGHT
TO FORAGE FOR THE
RICH BENTHIC
ALGAE THAT
COATS THE GRASS
AND OTHER
OBJECTS.

AFTER ABOUT A
YEAR, THEY'RE TWO
TO THREE INCHES LONG
AND NO LONGER BURY
THEMSELVES. IN TWO
YEARS THESE YOUNG
CONCHS WILL HAVE REACHED
FOUR TO SIX INCHES IN
LENGTH AND ARE CALLED "ROLLERS".
THE JUVENILES CONTINUE TO GROW
IN A SPIRAL FASHION AND ONLY
WHEN THREE TO FOUR YEARS OLD DO
THEY START TO BUILD THE BROAD,
FLARING, WINGLIKE OUTER LIP THAT IS
CHARACTERISTIC OF *STROMBUS GIGAS.*

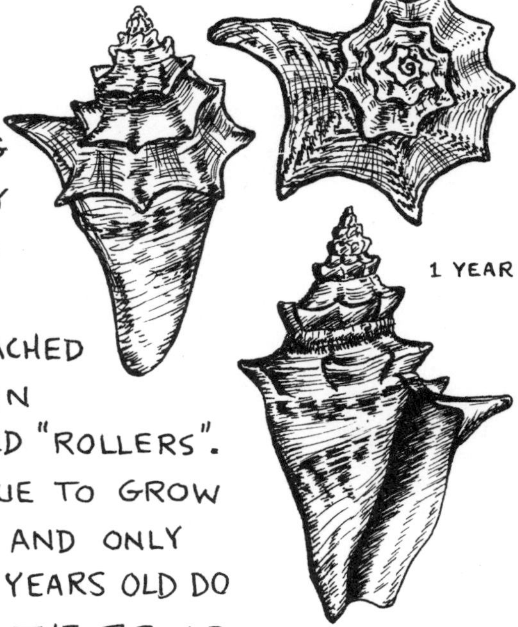

    IN JUST A FEW MONTHS THIS LIP IS
FULLY FORMED AND NOW THE
ANIMAL IS SEXUALLY
MATURE. IT HAS BECOME
A FAIR COPY OF THE
*STROMBUS
GIGAS*
THAT LAID
THE EGGS. THE
SHELL IS SO
BEAUTIFUL IT IS
SOUGHT AFTER
BY COLLECTORS
THE WORLD OVER;

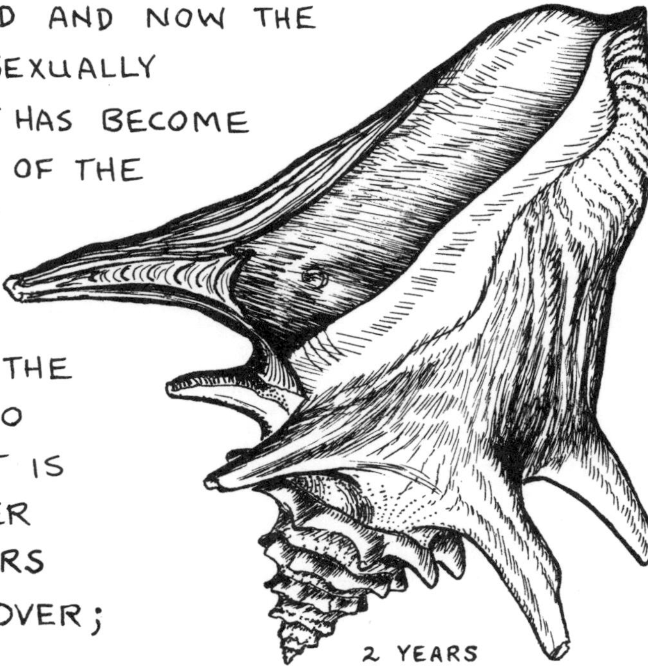

1 YEAR

2 YEARS

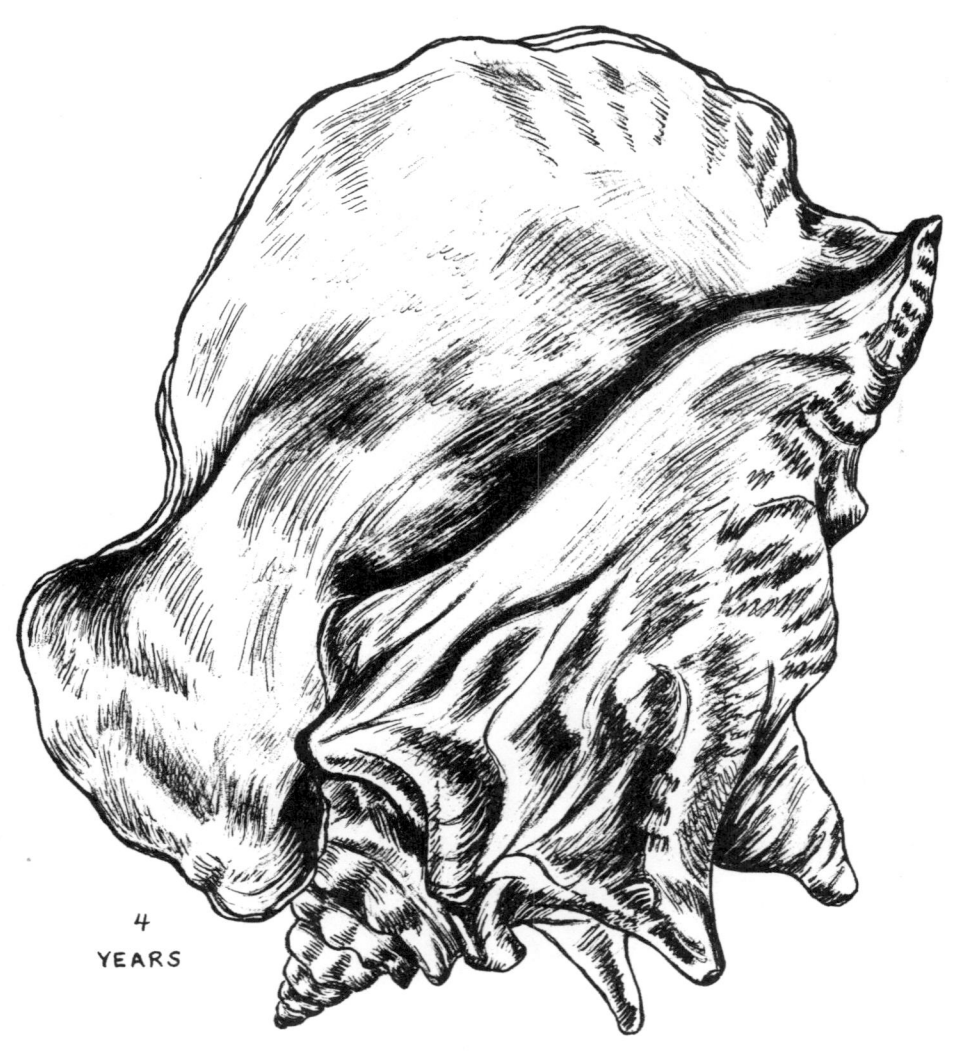

4
YEARS

THE MEAT HAS REACHED ITS MAXIMUM WEIGHT
AND IS PRIME FOR THE TABLE. THE SHELL MAY
MEASURE EIGHT TO TWELVE INCHES AND WEIGH
UP TO FIVE POUNDS.

      FROM HERE ON, HOWEVER, GROWTH SLOWS.
THE QUEEN GRADUALLY INCREASES THE OVERALL
THICKNESS OF THE SHELL, ADDING LAYER AFTER
LAYER OF CALCIUM CARBONATE. THE SPACE OC-
CUPIED BY THE SNAIL ACTUALLY BECOMES SMALLER.

AS TIME GOES BY THE OUTSIDE OF THE SHELL
SUFFERS FROM THE ACTION OF WORMS AND
CERTAIN BORING ORGANISMS. THE SPIRE
BECOMES WORN AND BLUNTED. AT THE
APERTURE THE MANTLE HAS ADDED PINK NACRE.

THESE OLD TIMERS WERE ONCE
THOUGHT TO BE A SEPARATE SPECIES AND
WERE MISTAKENLY NAMED *STROMBUS SAMBA*.
OLDER, HEAVY CONCHS ARE STILL REFERRED
TO AS "SAMBAS". THEIR FLESH TENDS TO
BE GRAYISH RATHER THAN WHITE AND MAY
BE BITTER, WITH POSSIBLE EMETIC EFFECTS.
THE MEAT, IF USED, SHOULD BE DOUBLE
BOILED.

CONCHS HAVE SOME REGENERATIVE
POWERS: EVEN A SAMBA, UPON LOSING
HER LIP, CAN REBUILD HERSELF A
FLARING BEAUTY LIKE A YOUNG SHELL.

THE LIFE SPAN
OF THE QUEEN CONCH
IS IN QUESTION.
SOME SAY SIX
TO TEN
YEARS AND
OTHERS
ASSERT THE
QUEEN WILL
LIVE FROM TEN TO
TWENTY-FIVE YEARS.

Beginnings

THE MISTS OF AGES SHROUD THE ORIGINS OF MOLLUSKS, BUT IT IS GENERALLY BELIEVED THEY APPEARED SOME 600 MILLION YEARS AGO. DURING THE CRETACEOUS PERIOD (BEGINNING 135 MILLION YEARS AGO) A RICH VARIETY OF THEM SHARED THE SEA WITH 12-FOOT TURTLES, 35-FOOT MARINE LIZARDS, AND 50-FOOT SHARKS. BY THE TERTIARY PERIOD (65 MILLION YEARS AGO) MOLLUSKS KNOWN AS GASTROPODS HAD EVOLVED TO SOMETHING LIKE THEIR PRESENT FORM. A GROUP OF GASTROPODS CALLED STROMBS FLOURISHED IN WARM, SHALLOW WATERS. ONE OF THESE, OUR *STROMBUS GIGAS*, CAPITALIZING ON A DIET OF SUN-FUELED ALGAE, DEVELOPED INTO ONE OF THE LARGEST OF ALL MARINE SNAILS — TEMPTING AND NOURISHING PREY FOR MANY CARNIVORES. IN RECENT TIMES MAN HAS PLAYED AN INCREASING ROLE IN THE QUEEN'S HISTORY: HE'S BEEN CRACKING CONCH SHELLS FOR THEIR MEAT SINCE HE FIRST ARRIVED IN THE CARIBBEAN BASIN A FEW THOUSAND YEARS AGO.

WHEREVER THE QUEEN LIVES — FROM BRAZIL
NORTHWARD THROUGHOUT THE WEST INDIES AND THE
BAHAMAS TO BERMUDA AND THE SHORES OF LOWER
FLORIDA AND THE KEYS — IT HAS BEEN A
MAJOR SOURCE OF PROTEIN FOR HUMANS.

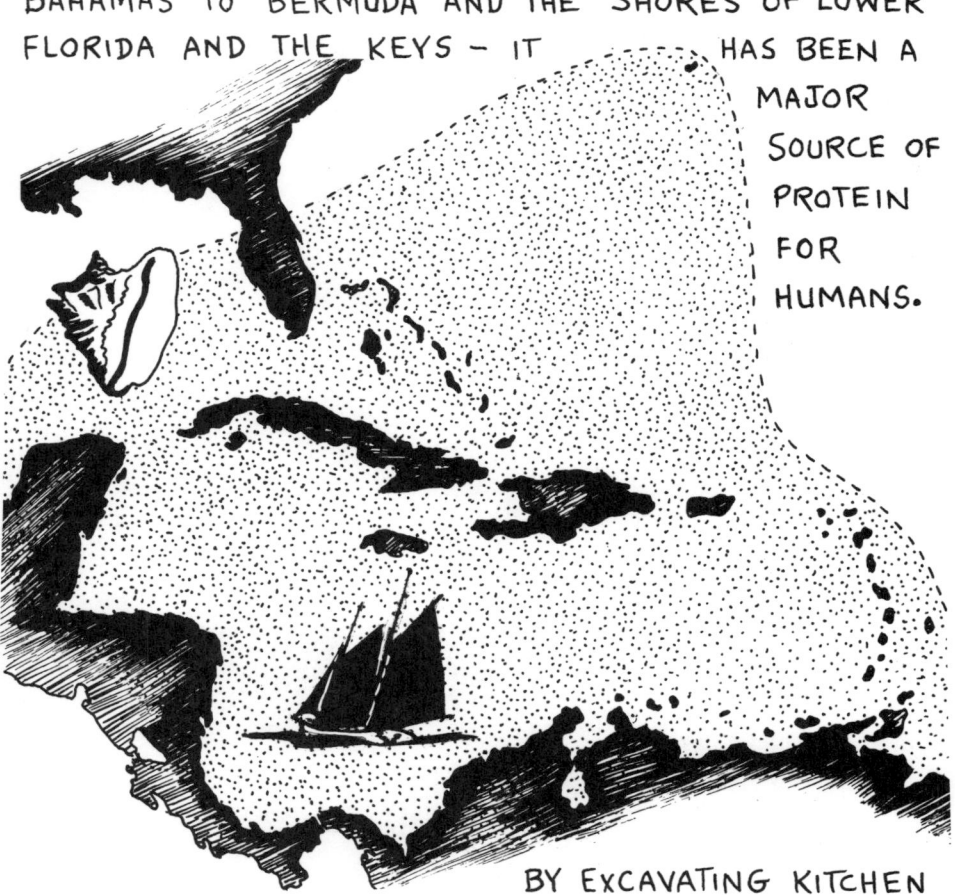

BY EXCAVATING KITCHEN
MIDDENS, ARCHAEOLOGISTS HAVE BEEN ABLE TO TRACE
THE INDIAN'S DEVELOPMENT FROM THE EARLIEST TIMES
AND HAVE FOUND JUST HOW
EXTENSIVELY THE CONCH WAS
USED FIRST FOR FOOD, THEN
AS A TOOL, AND LATER,
BECAUSE OF ITS BEAUTY,
FOR ADORNMENT AND TRADE.

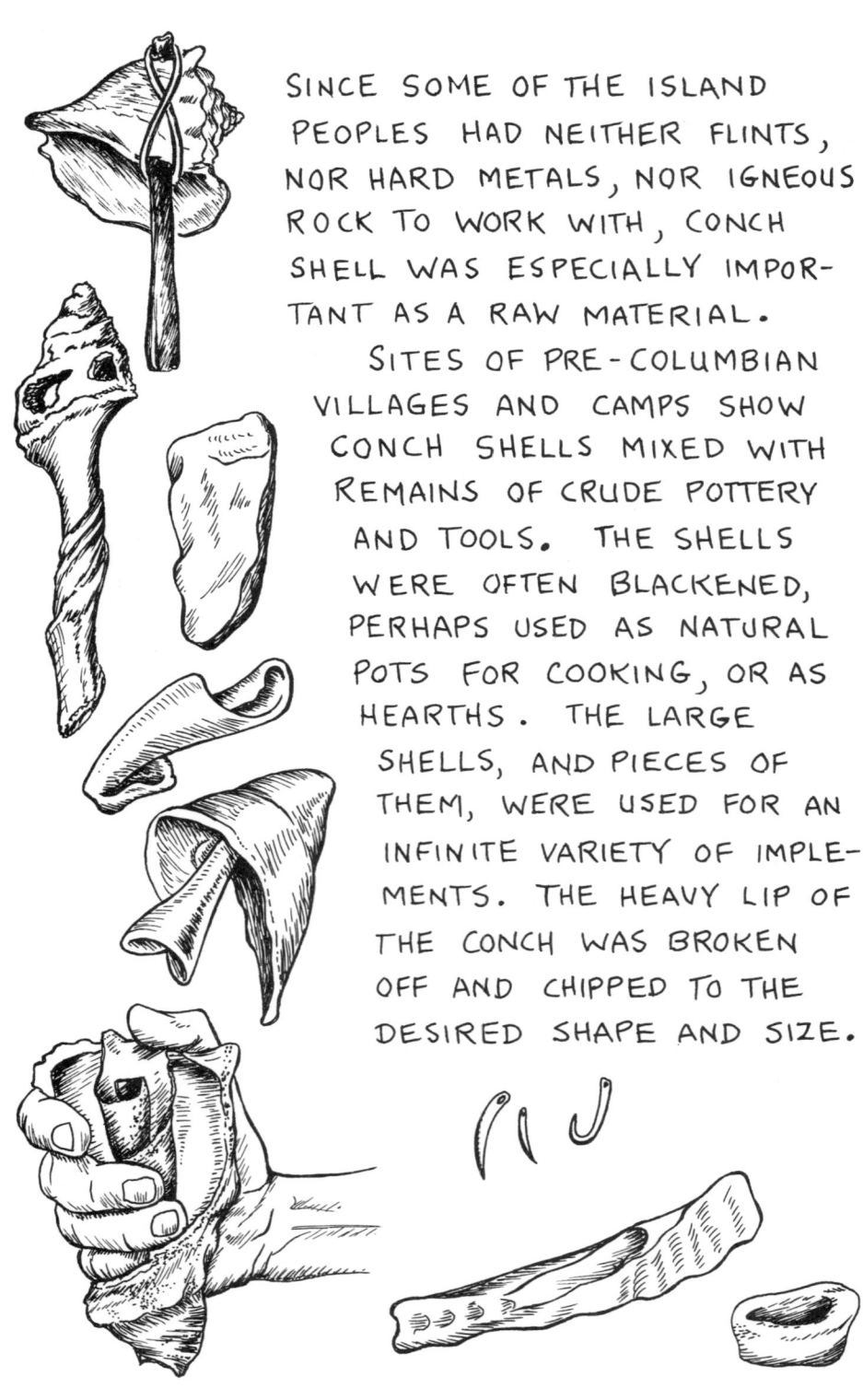

SINCE SOME OF THE ISLAND PEOPLES HAD NEITHER FLINTS, NOR HARD METALS, NOR IGNEOUS ROCK TO WORK WITH, CONCH SHELL WAS ESPECIALLY IMPORTANT AS A RAW MATERIAL.

SITES OF PRE-COLUMBIAN VILLAGES AND CAMPS SHOW CONCH SHELLS MIXED WITH REMAINS OF CRUDE POTTERY AND TOOLS. THE SHELLS WERE OFTEN BLACKENED, PERHAPS USED AS NATURAL POTS FOR COOKING, OR AS HEARTHS. THE LARGE SHELLS, AND PIECES OF THEM, WERE USED FOR AN INFINITE VARIETY OF IMPLEMENTS. THE HEAVY LIP OF THE CONCH WAS BROKEN OFF AND CHIPPED TO THE DESIRED SHAPE AND SIZE.

HAMMERS, PICKS, GOUGES,
CHISELS, ADZE BLADES,
SCRAPERS,
SCOOPS, DIPPERS,
CUPS, DISHES —
EVEN HEAVY FISHHOOKS
WERE FASHIONED FROM THE
SHELL OF *STROMBUS GIGAS*. CEREMONIAL
OBJECTS WERE MADE FROM IT. THE WHOLE SHELL
WAS USED AS A TRUMPET. ON THE ISLAND OF SAN
SALVADOR, A FEW ZEMIS (IDOLS) HAVE BEEN
FOUND MADE OF CONCH.

COLUMBUS MENTIONED IN NOTES ON HIS
FIRST AND THIRD VOYAGES THAT HE SAW
CANOES OF MAHOGANY, SOME OF WHICH WERE
SEVENTY-FIVE FEET LONG AND FIVE FEET
WIDE, CARRYING MORE THAN FIFTY PEOPLE.
THE TIMBER FOR THOSE CANOES WAS
CHARRED, THEN CHISELED TO REMOVE THE HUGE
AMOUNT OF WOOD TO MAKE THE WALLS THE
PROPER THICKNESS ..... SCRAPED OUT WITH
CONCH TOOLS? NO MEAN FEAT!

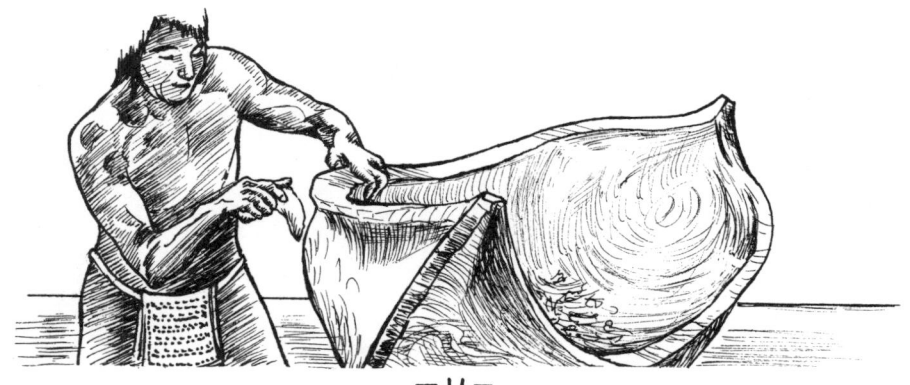

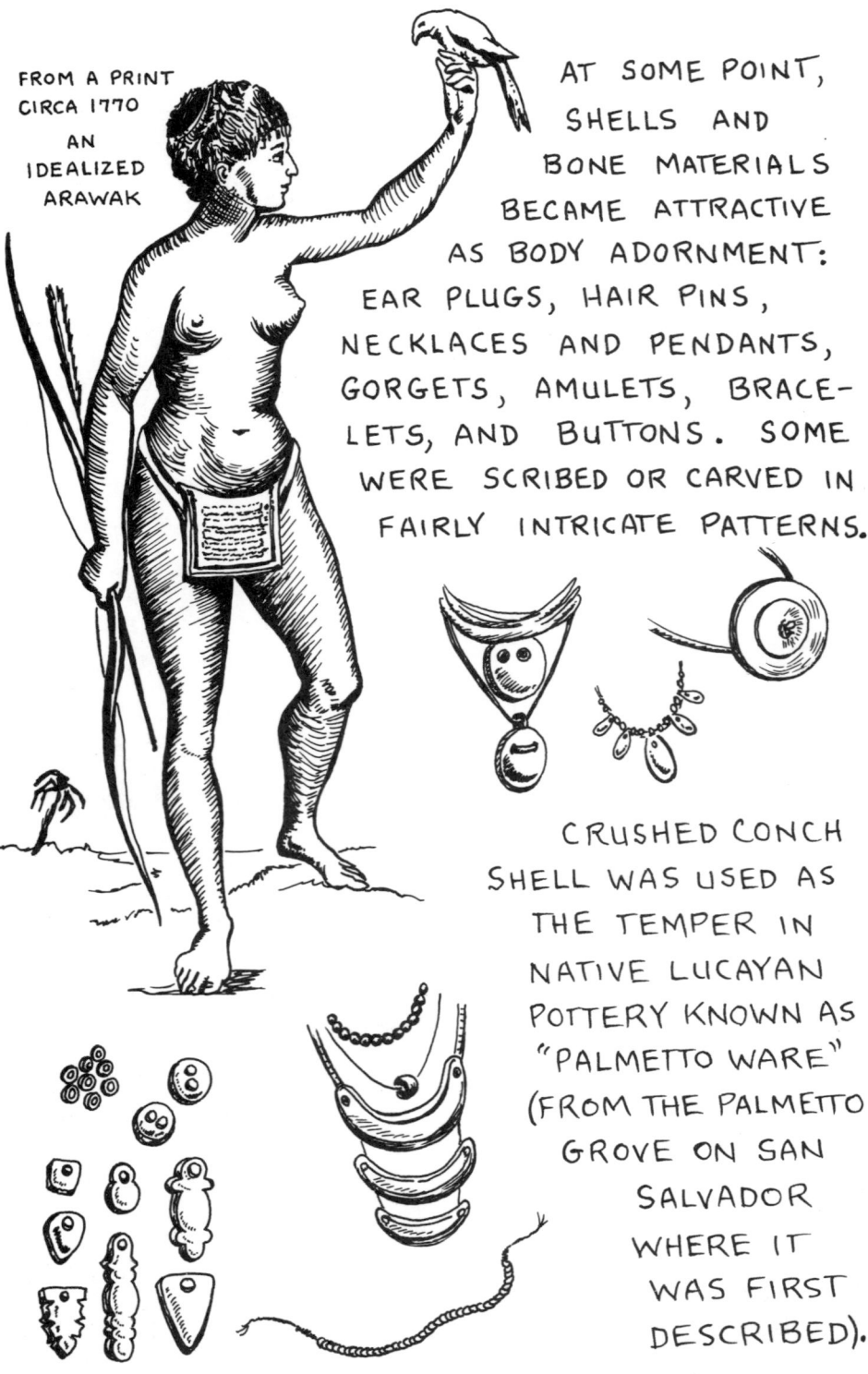

FROM A PRINT CIRCA 1770 AN IDEALIZED ARAWAK

AT SOME POINT, SHELLS AND BONE MATERIALS BECAME ATTRACTIVE AS BODY ADORNMENT: EAR PLUGS, HAIR PINS, NECKLACES AND PENDANTS, GORGETS, AMULETS, BRACELETS, AND BUTTONS. SOME WERE SCRIBED OR CARVED IN FAIRLY INTRICATE PATTERNS.

CRUSHED CONCH SHELL WAS USED AS THE TEMPER IN NATIVE LUCAYAN POTTERY KNOWN AS "PALMETTO WARE" (FROM THE PALMETTO GROVE ON SAN SALVADOR WHERE IT WAS FIRST DESCRIBED).

NIÑA

HISTORY
RELATES THAT
THE CONCH HELPED
SUSTAIN COLUMBUS
AND HIS
CREW. THE
JOURNAL OF HIS
SECOND VOYAGE
ON THE NIÑA
(1494) RECORDS
THAT ON THE SOUTH COAST OF CUBA CONCHS
"AS LARGE AS A CALF'S HEAD" COULD BE
SEEN ON THE BOTTOM. THE MEN COLLECTED
THEM BY THE BOATLOAD AND BOILED THE MEAT,
"AS BIG AS A MAN'S ARM", IN SEA WATER.

IT MAY BE
SAID THAT
COLUMBUS
DISCOVERED
AMERICA AND
THE QUEEN
CONCH AS
WELL.

FROM A WOODCUT
DONE IN 1493

SHORTLY AFTER THE FIRST SHIPS RETURNED FROM THE NEW WORLD, THE QUEEN CONCH BECAME PRIZED AS A SYMBOL OF THE EXOTIC, THE BEAUTIFUL, AND THE MYSTERIOUS.

TO SOME EXTENT, IT HAS REMAINED SO THROUGH THE CENTURIES. WOODCUTS DATING FROM THE SIXTEENTH CENTURY INCLUDE THE CONCH. BY QUEEN VICTORIA'S TIME, THE COLORFUL SHELL WAS ENSCONCED ON MANY A MANTLEPIECE. IT DECORATED GROTTOS AND FOUNTAINS IN THE ROCOCO STYLE. CONCH PEARLS WERE POPULAR IN THE SEVENTEENTH CENTURY. DURING THE EIGHTEEN HUNDREDS ENGLAND IMPORTED AS MANY AS 300,000 SHELLS A YEAR FOR USE IN THE MANUFACTURE OF FINE PORCELAIN. PAINTINGS BY CÉZANNE (LA PENDULE NOIRE) AND JAN DE HEEM (A STILL LIFE WITH PARROTS) PROMINENTLY INCLUDE THE BEAUTIFUL QUEEN.

IN THE 1960'S AND 70'S, SHELL LIPS WERE SAWN OFF AND EXPORTED FROM NASSAU TO ITALY TO BE CRUSHED AND USED IN "MOTHER-OF-PEARL" JEWELRY.

Conch~ology

PERHAPS WE SHOULD DEFINE THE WORD "CONCH". PRONOUNCED "KONK", THE TERM IS ACTUALLY GENERIC FOR ALL LARGE SPIRAL UNIVALVE MARINE SHELLS. IT COMES FROM THE GREEK **ΚΟΥΧη** (MEANING SHELL). HENCE CONCHOLOGY IS THAT BRANCH OF ZOOLOGY WHICH DEALS WITH SHELLS OR MOLLUSKS. MANY SHELLS AROUND THE WORLD ARE NICKNAMED "CONCHS", BUT IF YOU USE THE WORD IN THE BAHAMAS, SOUTH FLORIDA, OR THE WEST INDIES, IT WILL BE UNDERSTOOD TO MEAN *STROMBUS GIGAS.*

NOW, OUR FRIEND THE QUEEN CONCH IS A MOLLUSK. SHE HAS A FLESHY SNAIL BODY WITH A MUSCULAR FOOT THAT CARRIES A HORNY, SICKLE-SHAPED OPERCULUM OR BONY PLATE. HER LONG, NARROW, FLEXIBLE FOOT IS WELL SUITED FOR MOVING HER HEAVY SHELL.

SHE CAN PRY HERSELF FORWARD
BY EXTENDING HER FOOT AND
DIGGING HER CLAW-LIKE OPERCULUM
INTO THE SAND. SHE MOVES
ALONG IN A SERIES OF SLOW, JERKY
MOTIONS. SOME MOLLUSKS HAVE
AN OPERCULUM WITH WHICH THEY
CAN COMPLETELY CLOSE THEMSELVES
INTO THEIR SHELL. IT FORMS A TRAP DOOR.
THE CONCH'S "CLAW" IS NOT LARGE ENOUGH TO
COVER HER SHELL OPENING COMPLETELY. IT'S
USED PRIMARILY FOR LOCOMOTION. IT HAS BEEN
SUGGESTED THAT HER OPERCULUM MAY ALSO
SERVE AS A WEAPON TO FEND OFF PREDATORS.
THE QUEEN'S HEAD IS CROWNED
WITH TWO EYE STALKS FROM
WHICH BRANCH SENSORY
TENTACLES. THE LIGHT-
SENSITIVE EYES ON THE TIPS
OF THE STALKS HAVE COLOR-
FUL GOLDEN RINGS AND LOOK
RATHER LIKE AGATES. FROM BETWEEN THE
EYE STALKS PROTRUDES A MUSCULAR TUBE CALLED
THE PROBOSCIS, WHICH ENDS IN A MOUTH. THE
MOUTH CONTAINS THE UNIQUE MOLLUSCAN BAND
OF TEETH (RADULA) WITH WHICH
SHE SCRAPES OFF THE ALGAE   MOUTH
SHE
EATS.

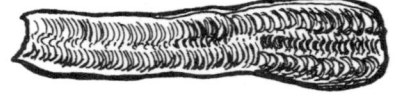

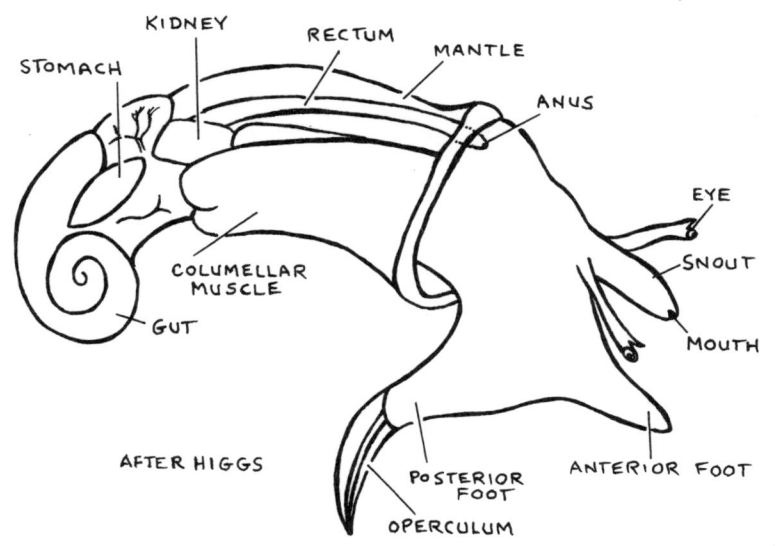

Diagram labels:
KIDNEY, STOMACH, RECTUM, MANTLE, ANUS, EYE, SNOUT, COLUMELLAR MUSCLE, GUT, MOUTH, AFTER HIGGS, POSTERIOR FOOT, ANTERIOR FOOT, OPERCULUM

AT THE BASE OF THE FOOT SPREADS THE MANTLE, WHICH IS BETWEEN THE HEAD AND THE INTERNAL ORGANS OF CIRCULATION, RESPIRATION, EXCRETION, DIGESTION, AND REPRODUCTION. THE FRONT PART OF THIS MANTLE EXTENDS TO FORM A SIPHON, THROUGH WHICH WATER IS DRAWN INTO THE MANTLE CAVITY TO THE FEATHERY GILLS HOUSED THERE. THE MANTLE (OR SKIRT) IS A THIN SHEET OF TISSUE WHICH SECRETES THE LIMY SHELL AND LAYS ON THE BEAUTIFUL PINK, PEACH, AND YELLOW COLORS THE CONCH IS SO FAMOUS FOR. THE MANTLE IS ALSO WHERE CONCH PEARLS ARE SOMETIMES FOUND.

A GELATINOUS PROTEIN ROD CALLED THE CRYSTALLINE STYLE IS HOUSED IN A SHEATH THAT TERMINATES IN THE SNAIL'S STOMACH. ITS END LIBERATES ENZYMES, WHICH HELP IN THE DIGESTION OF VEGETABLE MATTER. THE CONCH'S HEART HAS TWO CHAMBERS THAT CIRCULATE BLUEISH BLOOD, AS IS APPROPRIATE FOR ROYALTY. THE COLOR IS DUE TO A HIGH COPPER CONTENT, WHICH MAY HAVE CONTRIBUTED TO THE LOW INCIDENCE OF COPPER-DEFICIENCY ANEMIA AND POLIO-MYELITIS IN PEOPLES THAT CONSUME A LOT OF CONCH.

Foot

MANTLE

VERGE

OPERCULUM

FEMALE SHELLS ARE USUALLY LARGER THAN THOSE OF THE MALE. THE SEXES ARE SEPARATE AND FERTILIZATION IS INTERNAL. THE MALE COPULATORY ORGAN IS CALLED A "VERGE". (IF IT IS NIPPED OFF BY AN ERRING EEL OR AN UNCARING CRAB, THE MALE CONCH HAS THE ABILITY TO GROW HIMSELF A NEW ONE.)

Conch-enomics

Imagine, if you will, that far away time of the pre-columbian new world, when only the indians were around to note and profit from the conch's great abundance. Think of the millions of *STROMBUS*, rose-pink and peach and cream. Century after century they mated, bred and lived in the shallow seas. They died and were piled in windrows along the white beaches, glistening in the wash of the sea and the shine of the sun. They were ground by the surf into sand; some became embedded in limestone. Countless numbers were eaten and the shells added to the middens of indian villages.

Now civilized man, with great increase in numbers, great greed, and great disregard for the survival of a mere shellfish, has, in a few years' time, almost destroyed this most ancient and noble of mollusks.

It was probably inevitable – the large shell was so easy to catch – sitting in the grassy shallows. Why, a fellow could sometimes just pick up conch wading in knee-deep water. And if they were a little deeper, the water was so clear it was easy to dive them up. There they were, all year round and in water warm enough not to be a deterrent.

At first the queen was taken only for local consumption. Later sailing vessels carried them to trade with other islands.

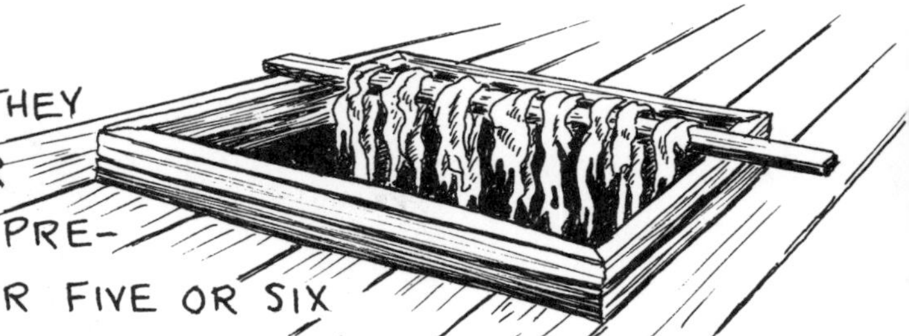

IN THOSE DAYS THEY
WERE DRIED FOR
SHIPPING, WHICH PRE-
SERVED THEM FOR FIVE OR SIX
WEEKS. * IN THE EARLY DAYS OF BAHAMIAN TRADE
WITH HAITI, DRIED CONCH WERE WORTH LESS THAN A
PENNY APIECE. HUNDREDS OF TONS WERE SHIPPED.

TODAY, CONCH MEAT LANDED FOR THE
BAHAMIAN MARKET IS WORTH
AS MUCH OR MORE
THAN GROUPER — UP TO
$3 A POUND IN NASSAU.
IN 1998 ALMOST 1½ MILLION
POUNDS OF MEAT WAS
HARVESTED THROUGH-
OUT THE BAHAMAS —
NOT COUNTING
CONCH TAKEN FOR

---

*

DRYING IS A SMELLY
BUSINESS. THE MEAT
IS HUNG FOR SEVERAL
DAYS IN THE SUN
UNTIL IT BECOMES
DESICATED. THIS
PRACTICE CONTINUES
TODAY IN THE FAR
OUT ISLANDS, WHERE
REFRIGERATION IS
LACKING.

PERSONAL USE BY RESIDENTS AND VISITORS. TO SUPPLY THE MARKET WITH SUCH RECORD AMOUNTS OF CONCH MEAT, FISHERMEN HAVE VIRTUALLY ABANDONED THE TRADITIONAL METHOD OF HOOKING WITH A LOOK BUCKET. THEY TRAVEL FURTHER AND USE HOOKAH (SURFACE COMPRESSOR) DIVING GEAR TO TAKE CONCH.*

THE RESULT IS PREDICTABLE: CONCH STOCKS ARE OVERFISHED IN 90% OF THEIR RANGE. HARVESTING AT PRESENT RATES, SOME OBSERVERS FORECAST, WILL BRING ABOUT THE COLLAPSE OF THE BAHAMIAN CONCH FISHERY IN LESS THAN A DECADE. SINCE 1991 THE COMMISSION ON INTERNATIONAL TRADE IN ENDANGERED

---

* THESE LARGE CATCHES RESULT IN MOUNTAINS OF SHELL CLOGGING HARBORS WHERE CONCH IS CLEANED AND THE MEAT SOLD ASHORE. IN NASSAU FINES DISCOURAGE DUMPING SHELL IN NAVIGABLE WATERS. FOR THE SAME REASON, ST. KITTS/NEVIS BANS LANDING CONCH SHELL AND ADMITS ONLY CLEANED MEAT.

SPECIES HAS LISTED THE QUEEN CONCH AS POTENTIALLY ENDANGERED.

IN MANY AREAS WHERE THE CONCH WAS FORMERLY A VIABLE FOOD RESOURCE IT IS PARTIALLY OR TOTALLY PROTECTED TO CONSERVE THE TINY POPULATIONS THAT REMAIN, BUT CONSERVATION LAWS ARE SOMETIMES AMBIVALENT AND/OR UNENFORCEABLE. SEVERAL CARIBBEAN ISLAND GOVERNMENTS HAVE ADOPTED CLOSED SEASONS. THE BAHAMIAN DEPARTMENT OF FISHERIES CONTROLS EXPORTS, PROHIBITS TAKING CONCH WITH SCUBA GEAR, AND BANS TAKING IMMATURE CONCH - DEFINED AS ANY CONCH "WHICH DOES NOT POSSESS A WELL FORMED FLARING LIP". FLORIDA HAS INSTITUTED A TOTAL BAN ON TAKING CONCH, WITH PENALTIES FOR POACHING. MARINE SANCTUARIES, WHERE TAKING ANY MARINE LIFE IS PROHIBITED, HAVE BEEN ESTABLISHED IN THE BAHAMAS AND SEVERAL CARIBBEAN NATIONS. IT IS HOPED THESE UNDERWATER NATIONAL PARKS WILL SERVE AS NURSERIES TO HELP REPLENISH SEAFOOD RESOURCES.

ALTHOUGH CONCH IS STILL A PRINCIPAL FOOD

SOURCE AND CONSTITUTES A PRIMARY FISHERY (ALONG WITH LOBSTER AND FINFISH) WHEREVER IT IS FAIRLY ABUNDANT, ITS CONTRIBUTION TO MARINE ECOLOGY IS FAR GREATER THAN ITS MARKET VALUE. CONCH IS A PRIMARY LINK IN THE FOOD CHAIN, NOURISHING AT SOME POINT IN ITS LIFE CYCLE NEARLY EVERYTHING THAT SWIMS. IN A SENSE, CONCHS ARE THE UNDER-WATER VERSION OF BEEF CATTLE - CONVERTING VEGE-TABLE DIET INTO MEAT PROTEIN FOR SPECIES HIGHER UP THE CHAIN.

PERHAPS THE BEST HOPE FOR THE FUTURE OF THE BELEAGUERED QUEEN LIES IN THE DRAMATIC AD-VANCE OF CONCH MARICULTURE. UNTIL VERY RECENTLY, LITTLE WAS KNOWN ABOUT THE LIFE HISTORY OF THE CONCH, EVEN BY THE FISHERMEN WHOSE LIVES ARE STRUCTURED AROUND IT. IN 1980 RESEARCHERS AT THE UNIVERSITY OF MIAMI PAVED THE WAY TO RAISING BABY CONCHS FROM EGGS COLLECTED IN THE WILD, MAKING THE POSSIBILITY OF "CONCH FARMS" A REALITY. EXPERIMENTS IN VENEZUELA, THE TURKS AND CAICOS ISLANDS, BAHAMAS, MEXICO, PUERTO RICO, CUBA, AND FLORIDA HAVE PROVED THE VIABILITY OF THE TECHNIQUE. A NOTABLE SUCCESS IS THE CAICOS CONCH FARM,

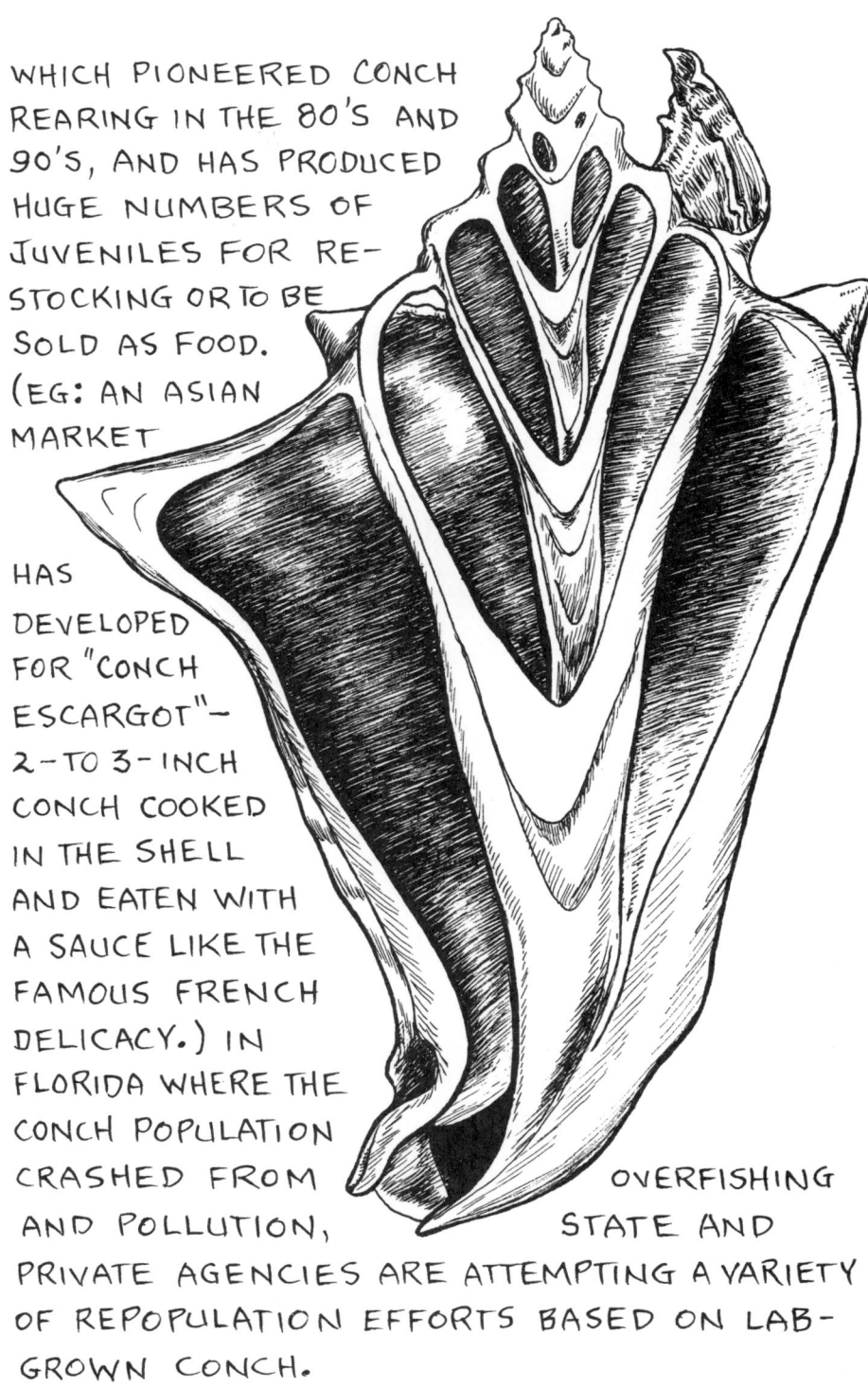

WHICH PIONEERED CONCH
REARING IN THE 80'S AND
90'S, AND HAS PRODUCED
HUGE NUMBERS OF
JUVENILES FOR RE-
STOCKING OR TO BE
SOLD AS FOOD.
(EG: AN ASIAN
MARKET

HAS
DEVELOPED
FOR "CONCH
ESCARGOT"—
2-TO 3-INCH
CONCH COOKED
IN THE SHELL
AND EATEN WITH
A SAUCE LIKE THE
FAMOUS FRENCH
DELICACY.) IN
FLORIDA WHERE THE
CONCH POPULATION
CRASHED FROM   OVERFISHING
AND POLLUTION,   STATE AND
PRIVATE AGENCIES ARE ATTEMPTING A VARIETY
OF REPOPULATION EFFORTS BASED ON LAB-
GROWN CONCH.

Until very recent times, the conch fishery in the Bahamas employed some of the last boats in the western hemisphere to work commercially under sail. Usually hand-built on the shores of their home island, these unique

and picturesque craft (called "smacks") towed fourteen- to eighteen-foot open dinghies to the conching grounds. The dinghies, with a two-man crew, were sailed or sculled from the smack to the conch beds where the shellfish were taken by "hooking".

The tools of this trade are a glass-bottomed bucket and a "conch hook".

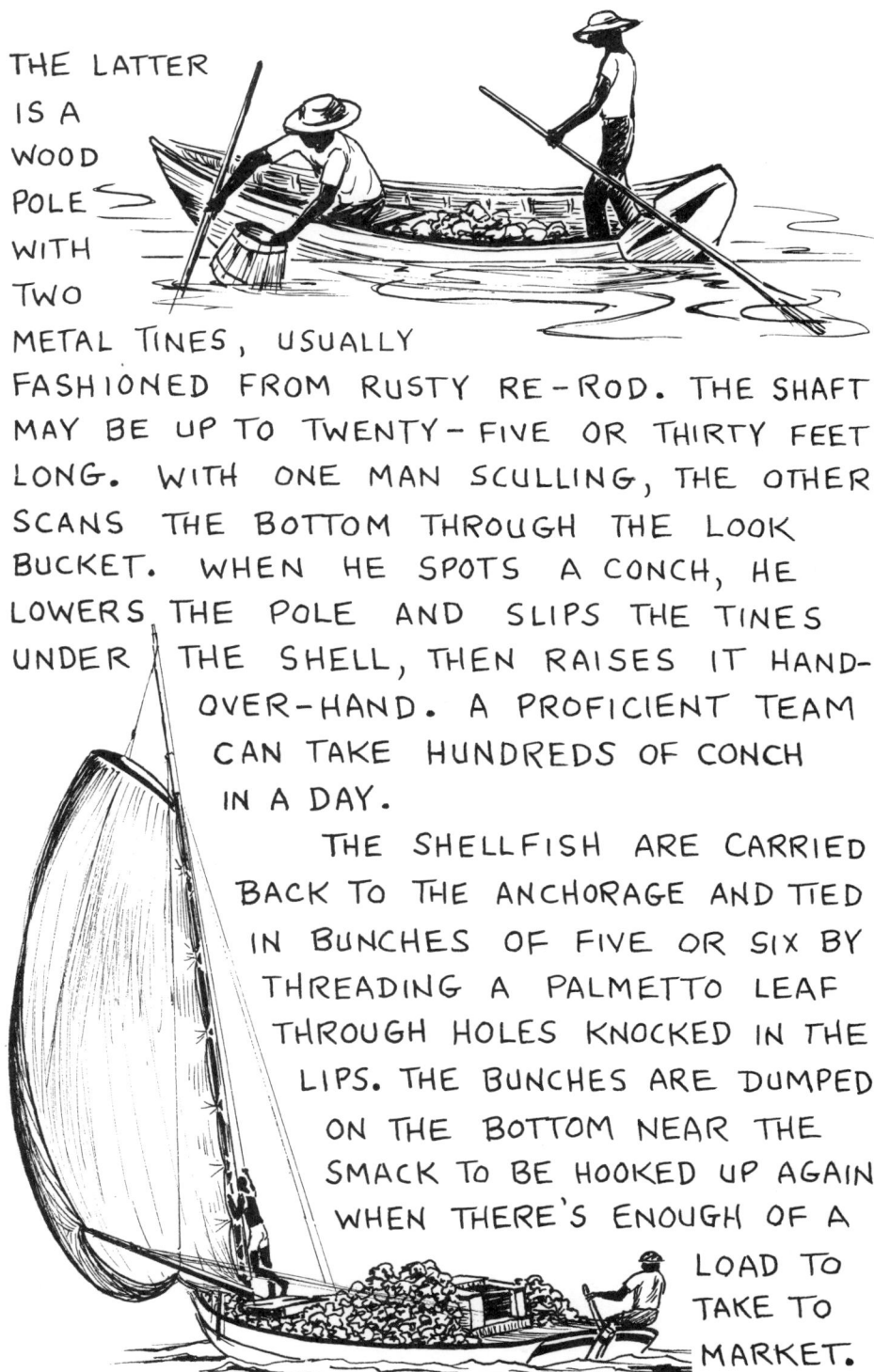

THE LATTER IS A WOOD POLE WITH TWO METAL TINES, USUALLY FASHIONED FROM RUSTY RE-ROD. THE SHAFT MAY BE UP TO TWENTY-FIVE OR THIRTY FEET LONG. WITH ONE MAN SCULLING, THE OTHER SCANS THE BOTTOM THROUGH THE LOOK BUCKET. WHEN HE SPOTS A CONCH, HE LOWERS THE POLE AND SLIPS THE TINES UNDER THE SHELL, THEN RAISES IT HAND-OVER-HAND. A PROFICIENT TEAM CAN TAKE HUNDREDS OF CONCH IN A DAY.

THE SHELLFISH ARE CARRIED BACK TO THE ANCHORAGE AND TIED IN BUNCHES OF FIVE OR SIX BY THREADING A PALMETTO LEAF THROUGH HOLES KNOCKED IN THE LIPS. THE BUNCHES ARE DUMPED ON THE BOTTOM NEAR THE SMACK TO BE HOOKED UP AGAIN WHEN THERE'S ENOUGH OF A LOAD TO TAKE TO MARKET.

THEN THEY ARE
PILED ON THE DECK AND SAIL IS SET FOR
NASSAU OR SOME OTHER MARKET TOWN. DE-
PENDING ON WEATHER, THIS TRIP MAY TAKE
OVERNIGHT OR SEVERAL DAYS. REPEATED
WETTINGS KEEP THE CONCH ALIVE AND
HEALTHY. WHEN THE SMACK PUTS INTO
PORT, THE BUNCHES OF CONCH ARE AGAIN
DROPPED IN THE SEA ALONGSIDE,
TO BE HOOKED
UP AND
CLEANED
AS THE
MARKET
DEMANDS.

the Queen's Court

THE QUEEN DOESN'T ALWAYS LIVE ALONE.
THERE ARE A FEW LITTLE
FELLOWS WHO CONSORT
WITH HER AND BENEFIT
FROM THE ASSOCIATION.
THAT'S CALLED COMMENSALISM,
MEANING THEY SHARE THE SAME
TABLE AND ARE HARMLESS TO EACH OTHER.
A VERY SMALL SPECIES OF CARDINAL FISH
(*APOGON STELLATUS*), TWO INCHES OR LESS,
USES THE MANTLE CAVITY OF THE CONCH AS A
RETREAT. COMMONLY KNOWN AS A
CONCH FISH, IT EATS INVERTEBRATES.
TINY SHRIMP COHABIT AND
LITTLE CRABS (*PORCELLANA
SAYANA*) MAY SCURRY OUT FROM
THE SNAIL'S MANTLE. LIMPETS
SOMETIMES CLING TO
THE OUTSIDE OF AN
ADULT SHELL, AND
CORALINE GROWTHS
MAY ADORN A
SAMBA.

the Queen's Enemies

*O*F ALL THE ENEMIES OF THE QUEEN, THE FIRST AND FOREMOST IS MAN *!* OTHERS THAT PREY UPON HER AND LUST AFTER HER FLESH: THE SPOTTED EAGLE RAY, WEIGHING UP TO FIVE HUNDRED POUNDS, HAS FLATTENED PLATES IN HIS JAW WITH WHICH HE CRUSHES STROMBUS. THESE DRAMATIC AND BEAUTIFUL RAYS ARE SADLY BECOMING RARE — A DECLINE DUE IN PART, NO DOUBT — TO THE INCREASING RARITY OF A FAVORITE FOOD.

THE LOGGERHEAD TURTLE IS ALSO CAPABLE OF CRUSHING THE HEAVY CONCH SHELL AND IS RESPONSIBLE FOR MANY OF THE BROKEN FRAGMENTS FOUND IN CONCH BEDS.

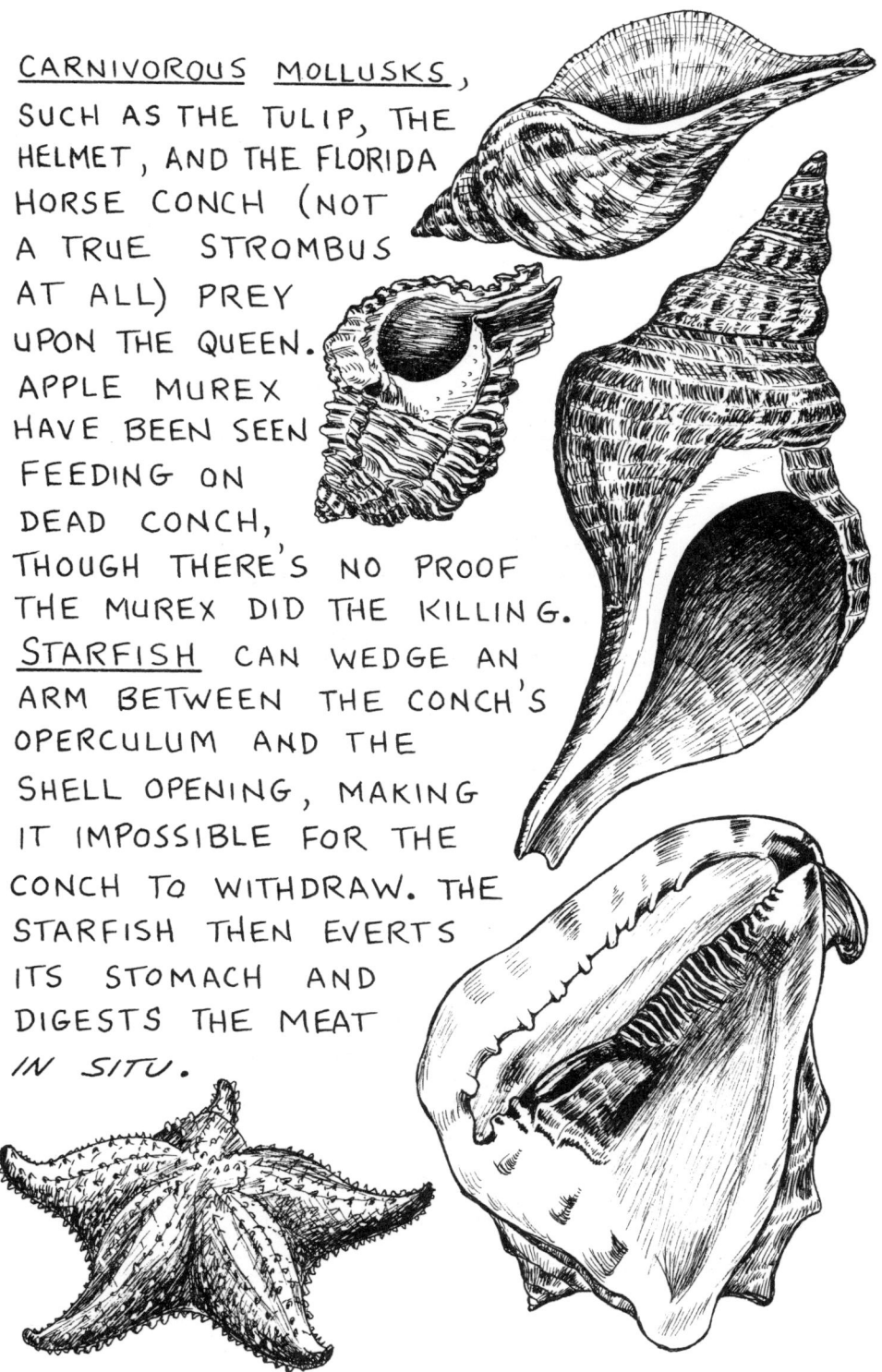

CARNIVOROUS MOLLUSKS, SUCH AS THE TULIP, THE HELMET, AND THE FLORIDA HORSE CONCH (NOT A TRUE STROMBUS AT ALL) PREY UPON THE QUEEN. APPLE MUREX HAVE BEEN SEEN FEEDING ON DEAD CONCH, THOUGH THERE'S NO PROOF THE MUREX DID THE KILLING. STARFISH CAN WEDGE AN ARM BETWEEN THE CONCH'S OPERCULUM AND THE SHELL OPENING, MAKING IT IMPOSSIBLE FOR THE CONCH TO WITHDRAW. THE STARFISH THEN EVERTS ITS STOMACH AND DIGESTS THE MEAT IN SITU.

HERMIT CRABS ADOPT EMPTY MARINE SHELLS TO HOUSE THEIR LONG, SPIRALLY-COILED SOFT ABDOMEN. THESE CRABS ATTACK LIVE CONCH BY INVERTING THE SHELL AND INSERTING BOTH POWERFUL CLAWS INTO THE OPENING.

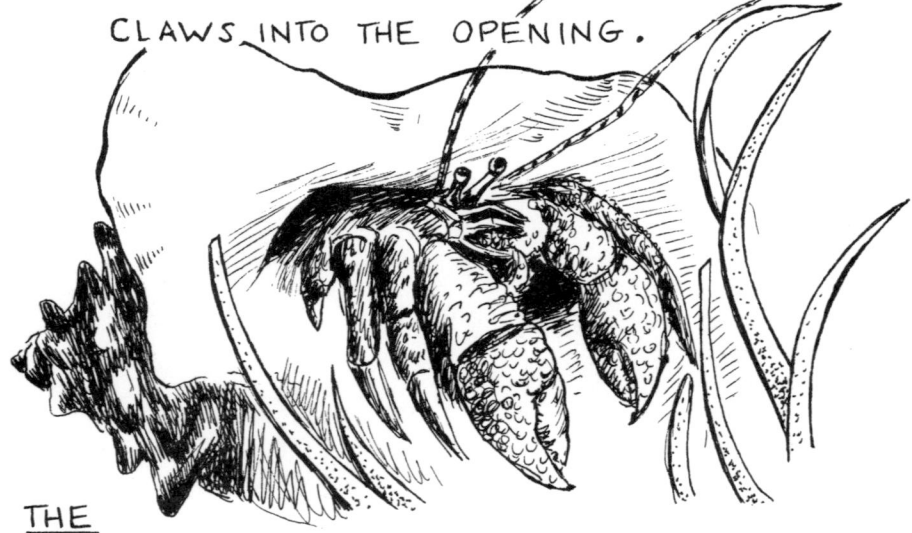

THE
OCTOPUS HAS BEEN OBSERVED WITH HIS ARMS AROUND A ROLLER CONCH, AND IT IS CERTAIN THE EMBRACE WAS NOT ROMANTIC. MANY'S THE OCTOPODAL DEN THAT IS CLUTTERED WITH EMPTY CONCH SHELLS. SOMETIMES A SMALL OCTOPUS WILL USE AN EMPTY CONCH SHELL AS A HOUSE.

FISH LIKE THE HOGFISH,
QUEEN
TRIGGER-
FISH,

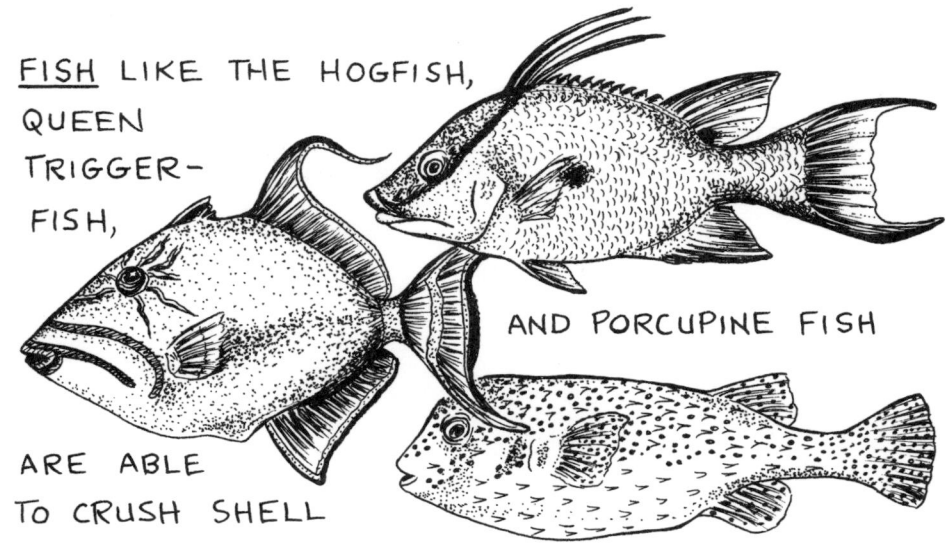

AND PORCUPINE FISH

ARE ABLE
TO CRUSH SHELL
EITHER WITH THEIR JAWS OR PHARYNGEAL TEETH
(A SECOND SET OF TEETH IN THE THROAT
REGION). THEY TAKE ONLY VERY YOUNG ROLLERS.

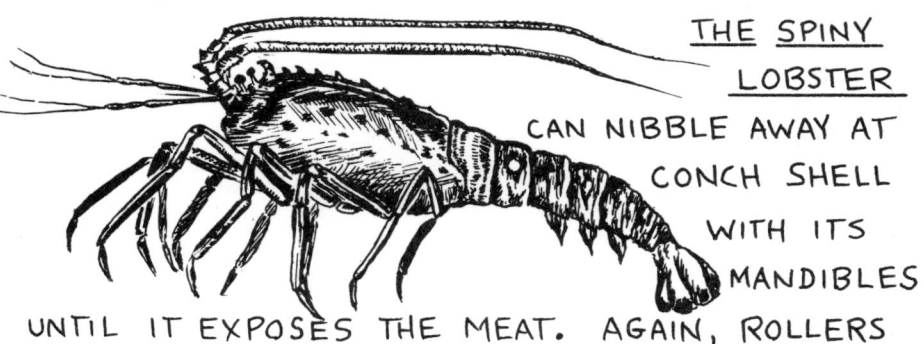

THE SPINY
LOBSTER
CAN NIBBLE AWAY AT
CONCH SHELL
WITH ITS
MANDIBLES
UNTIL IT EXPOSES THE MEAT. AGAIN, ROLLERS
ARE THE TARGET.

NURSE AND OTHER SHARKS HAVE A
FONDNESS FOR ROLLERS. ENTIRE CONCH
SHELLS HAVE BEEN FOUND
IN THEIR
STOMACHS.

the Queen's family

GOLIATH

SIX OTHER MEMBERS
OF THE CONCH FAMILY
(STROMBIDAE) SHARE
THE QUEEN'S DOMAIN.
THEY ALL ARE
HERBIVORES. ALL
HAVE A HORNY,
CURVED OPERCULUM.
NONE ARE SOUGHT
FOR FOOD.

STROMBUS GOLIATH
IS THE LARGEST
STROMBUS IN THE
WORLD. IT'S RARE AND
COVETED AND LIVES OFF
THE COAST OF BRAZIL. SOME
SPECIMENS BOUGHT IN THE NATIVE
MARKET HAVE A LENGTH OF FIFTEEN INCHES.
STROMBUS PUGILIS, THE FIGHTING CONCH, IS A
YELLOWISH- BROWN SHELL WITH A BRILLIANT
ORANGE APERTURE.

3-4"

PUGILIS

STROMBUS ALATUS, THE FLORIDA STROMB, IS VERY
SIMILAR IN SIZE AND LOOKS TO THE FIGHTING
CONCH, BUT LACKS THE SHOULDER SPINES (EXCEPT
IN THE LAST WHORL). ALSO, THE APERTURE IS
COMMONLY DARK BROWN.

STROMBUS
COSTATUS,
THE MILK
CONCH,
HAS A
CREAMY
AND
WHITE
COLORA-
TION. THE
OUTER LIP
IN OLDER
INDIVIDUALS
MAY
HAVE A
NOTICEABLE
ALUMINUM-
LIKE
GLAZE.

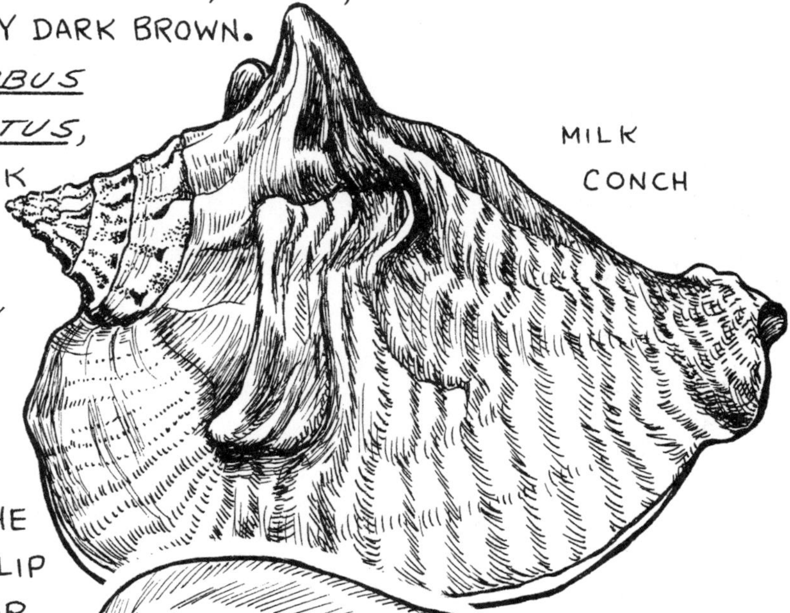

MILK
CONCH

4-6"

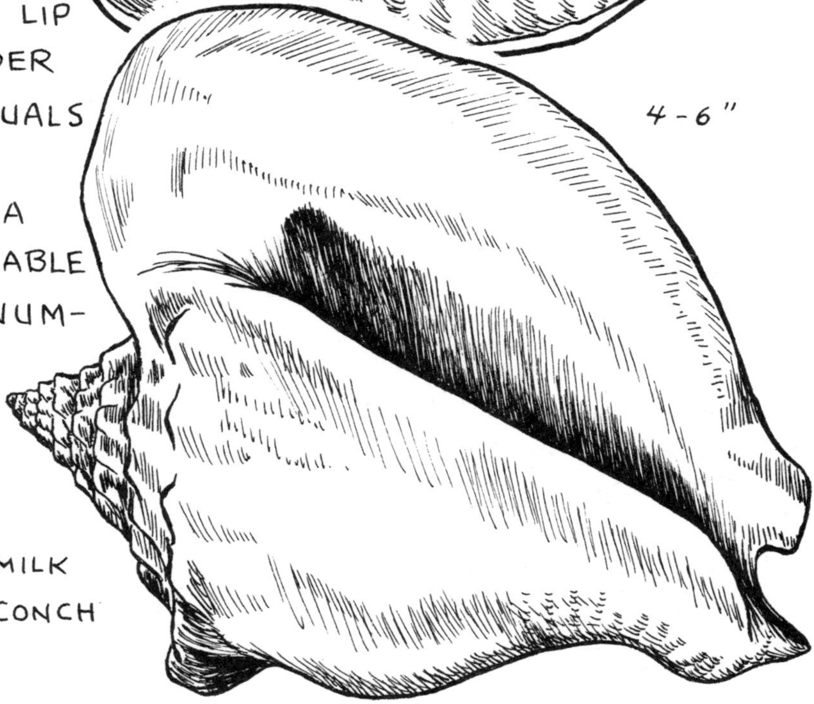

MILK
CONCH

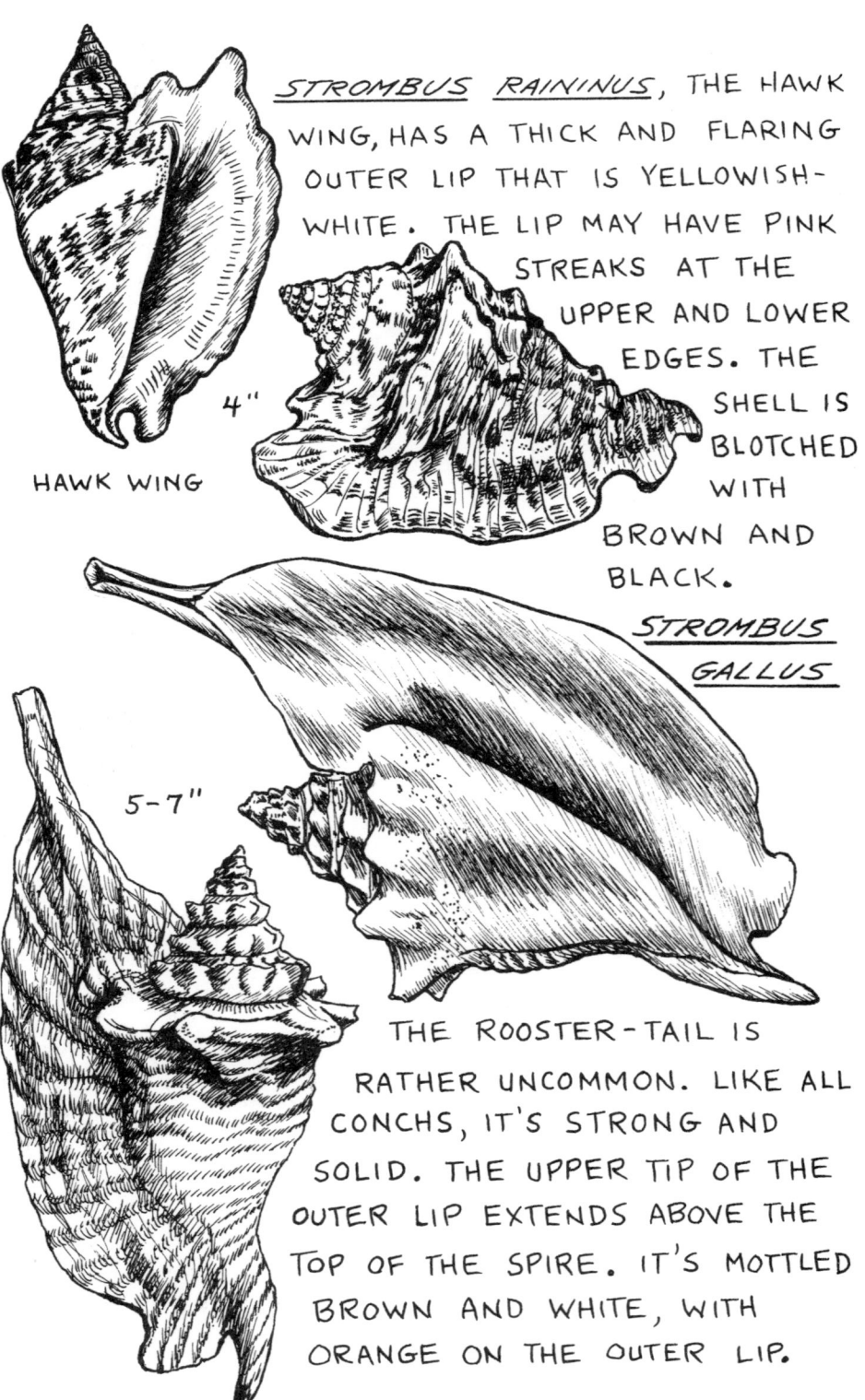

STROMBUS RAININUS, THE HAWK
WING, HAS A THICK AND FLARING
OUTER LIP THAT IS YELLOWISH-
WHITE. THE LIP MAY HAVE PINK
STREAKS AT THE
UPPER AND LOWER
EDGES. THE
SHELL IS
BLOTCHED
WITH
BROWN AND
BLACK.

STROMBUS
GALLUS

4"

HAWK WING

5-7"

THE ROOSTER-TAIL IS
RATHER UNCOMMON. LIKE ALL
CONCHS, IT'S STRONG AND
SOLID. THE UPPER TIP OF THE
OUTER LIP EXTENDS ABOVE THE
TOP OF THE SPIRE. IT'S MOTTLED
BROWN AND WHITE, WITH
ORANGE ON THE OUTER LIP.

the Queen's jewels

As everyone knows, a pearl is formed as an abnormal growth within the shell of a mollusk. It's secreted by the mantle as a protection against some foreign body: a parasite or a grain of sand. So it is with the queen and layers of the same nacre that color the lip are built up around the offending object. Conch pearls are rare and may become as large as a grape.

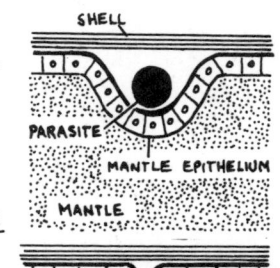
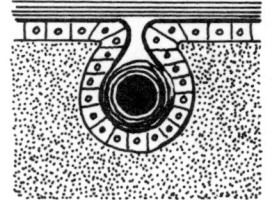
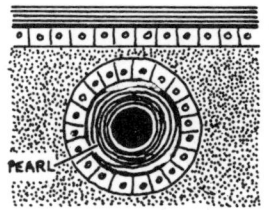

(AFTER HAAS)

Conch pearls bring excellent prices on the world market. Specimens cost as much as $300 at the 1880 Great International Fisheries Exhibition in London. A fine individual once fetched $5000 in Nassau. Current values range up to $1000.

The bright coloration of the pearls fades in time and they lose their luster. Fading is accelerated by exposure to direct sunlight.... Just as the queen herself will fade sitting on a windowsill.

Cameos are often carved in conch shell and the deep pink central column of the shell is cut and polished to make jewelry.

CONCH Curiosities

A <u>CONCH</u> <u>CRAWL</u> IS NOT A KIND OF WEST INDIAN DANCE. IT IS A SHALLOW AREA WALLED IN WITH EITHER OLD CONCH SHELL, CORAL OR DEBRIS WHERE CAPTURED CONCH ARE KEPT ALIVE FOR LATER USE. FROM THE DUTCH "KRAAL" MEANING PEN.

<u>SCORCHED</u> <u>CONCH</u> IS THE CLEANED MUSCLE SCORED IN A CRISSCROSS PATTERN TO TENDERIZE AND OFFER MORE SURFACE TO SEASONINGS AND MARINADES. A CORRUPTION OF "SCOTCHED", FROM THE ANGLO - FRENCH WORD "ESCOCHER" — TO CUT SUPERFICIALLY. IT IS NOT BURNED CONCH MEAT.

<u>ARE</u> <u>CONCHS</u> <u>RIGHT</u> <u>OR</u> <u>LEFT</u> <u>HANDED</u>? LIKE MOST GASTRO-PODS, CONCHS ARE DEXTRAL, THAT IS, THEIR SPIRAL SHELLS GROW IN A CLOCK-WISE DIRECTION (WHEN VIEWED FROM THE SPIRE).

WHEN THE QUEEN IS BARELY HATCHED, A CURIOUS THING HAPPENS TO HER. HER HEAD

SHELL
ANUS
ANUS
FOOT
MOUTH
FOOT
MOUTH
(AFTER PATTEN & ROBERT)

AND FOOT REMAIN STATIONARY, WHILE HER INTERNAL ORGANS ROTATE THROUGH AN ANGLE OF 180°. HER SOFT VISCERA IS PUSHED UP INTO A HUMP AND HER ANUS COMES TO REST ABOVE HER MOUTH INSTEAD OF AT THE OTHER END. (WHAT ELSE CAN YOU DO IF YOUR HOUSE HAS ONLY ONE OPENING!) THE ORGANS ON ONE SIDE OF HER BODY FAIL TO DEVELOP, AND HER VISCERAL MASS AND MANTLE (AND HER SHELL) BECOME SPIRALLY COILED. ONCE IN A GREAT RARE WHILE, A BABY WILL GET CONFUSED AND TWIST THE WRONG WAY, CREATING A SINISTRAL OR LEFT HANDED SHELL. SHE'D BE WHAT SHELL COLLECTORS CALL A "SPORT", AND GREATLY IN DEMAND. A FEW SHELLS, E.G. THE PERVERSE WHELK AND LIGHTNING WHELK, ARE NORMALLY SINISTRALLY COILED.

CONCHORCHESTRA? YES, THEY ACTUALLY HAVE A CONCH BLOWING CONTEST IN KEY WEST ONCE A YEAR AND THOUGH IT'S HARD TO BELIEVE, WHOLE TUNES, INCLUDING CLASSICAL PIECES, HAVE BEEN BLOWN.

THE CRYSTALLINE STYLE IS NOT A NEW CLOTHING FAD. NATIVES UP AND DOWN THE CHAIN OF THE WEST INDIES AVIDLY SLURP DOWN THIS PROTEIN ROD FROM THE CONCH'S STOMACH AND WITH A TWINKLE (OR A LEER) LET YOU KNOW — "THAT'LL MAKE YOU A MON!" IN FACT, CONCH— RAW, COOKED OR PICKLED— HAS ALWAYS BEEN TOUTED AS AN APHRODISIAC.

IS THERE A KING CONCH? NO, BUT SOMETIMES *CASIS TUBEROSA* (THE KING HELMET) IS MISTAKENLY CALLED KING CONCH — AND MIS- TAKENLY EATEN FOR CONCH AS WELL. HIS FLAVOR IS BITTER. HE'S ACTUALLY NO RELATIVE OF *STROMBUS GIGAS* AT ALL.

HOW MANY CONCH MAKE A BUNCH? A BAHAMIAN, WHEN ASKED WHY CONCH WERE LEFT IN THE SHALLOWS TIED IN BUNCHES OF FIVE OR SIX, SAID "TWO OR THREE, THAT MIGHT WALK OFF, BUT FIVE—THAT CAN NEVER DECIDE WHICH WAY TO GO!"

ARE CONCHS PEOPLE? YES, IF THEY LIVE IN KEY WEST OR THE BAHAMAS. THEY DON'T HAVE TO BE ROYALTY OR HAVE HARD SHELLS. THE FIRST HUMAN "CONCHS" WERE TORY SYMPATHIZERS WHO MOVED TO THE BAHAMAS TO ESCAPE THE AMERICAN REVOLUTION. THEY EARNED THE NAME WHEN THEY SAID THEY'D RATHER EAT CONCH THAN GO TO WAR. WHEN DESCENDENTS OF THOSE BAHAMIANS MOVED TO KEY WEST IN THE 1880'S, THEY CARRIED THE LABEL WITH THEM. PRESENT DAY NATIVE KEY WESTERS STILL TAKE PRIDE IN BEING CALLED "CONCHS". "CONKIE JOE" IS A TERM USED FOR A WATERFRONT TYPE.

WHAT IS HURRICANE HAM? IT'S WHAT DRIED CONCH WAS KNOWN AS WHEN OTHER MEATS AND FISH WERE IN SCARCE SUPPLY AFTER HURRICANES.

WHAT'S LAMBIE? IF MARY HAD A LITTLE LAMBIE, IT WAS A CONCH, BECAUSE THAT'S WHAT CONCH IS CALLED IN THE WEST INDIES. PERHAPS IT IS SHORT FOR "HURRICANE LAMB" OR DERIVED FROM THE LATIN NAME OF SOME PACIFIC CONCHS - *LAMBIDAE.*

<u>CONCH HOUSES</u>? THE FIRST ISLAND SETTLERS LEARNED TO REDUCE CONCH SHELLS BY BURNING TO OBTAIN LIME, FROM WHICH THEY MADE MORTAR BY ADDING SAND AND WATER. SINCE THEY HAD NO BRICKS OR FIELD STONE, THE STURDY CONCH SHELL WAS ITSELF OFTEN CEMENTED INTO WALLS AND FLOORS.

THE NAME "CONCH" AND THE SHELL AS A SYMBOL ARE USED EXTENSIVELY AROUND THE FLORIDA KEYS AND THROUGHOUT THE BAHAMAS AND WEST INDIES, VIZ: FIGHTING CONCHS AND CONCHETTES ARE SCHOOL TEAMS. THERE'S THE TOURIST CONCH TRAIN, THE CONCH SHELL NEWSPAPER (IT COVERS THE FLORIDA KEYS), THE CONCH INN, AND THE CONCH CRAWL (BAHAMIAN HOSTELRIES) AND THE PINK PEARL (A BAR). CONCHS ADORN BAHAMIAN PAPER MONEY AND COINS. T SHIRTS SAY "CONKED OUT" AND CONCH PEARL DRINKS ARE "IN". THE NAME IS GIVEN TO EMPORIUMS AND TOURIST TRAPS AD INFINITUM.

To Clean the Queen

IF YOU WATCH A NATIVE EXPERT ON THE WATERFRONT OPEN AND CLEAN CONCH, IT LOOKS LIKE A CINCH, BUT MANY A NEOPHYTE HAS GIVEN UP IN DISGUST OR FRUSTRATION WHEN FACED WITH THE HORNY BEAST. TO SAVE YOU FROM BLOWING YOUR TOP, AND THAT OF THE CONCH SHELL AS WELL, FOLLOW THE TECHNIQUES ILLUSTRATED HERE. CLEAN YOUR FIRST CONCH METHODICALLY. STUDY THE ANATOMY AS YOU GO, AND FUTURE ENCOUNTERS WILL BE EASIER ON BOTH OF YOU. FOR TOOLS YOU'LL NEED:

1) A MASON'S HAMMER, OLD HATCHET, CLAW HAMMER, OR A SIMILAR TOOL. (FOR "KNOCKING")

2) A SHARP, THIN-BLADED KNIFE.

3) A PAIR OF PLIERS. (UNLESS YOUR HANDS ARE VERY STRONG)

A. PICK UP YOUR CONCH IN THE LEFT HAND, WITH THE APERTURE DOWN. THE SPIRE WILL BE TOWARDS YOU AND YOU'LL BE HOLDING THE BODY OF THE SHELL. IF YOU HOLD THE SHARP LIP, YOU MAY BE INVITING A CUT. STRIKE THE INDICATED PLACE ON THE SPIRE, BETWEEN THE SECOND AND THIRD ROWS OF NODES — UNTIL A CRESCENT-SHAPED HOLE (OR SLIT) HAS BEEN MADE. THIS IS "KNOCKING".

B. THIS HOLE GIVES ACCESS TO THE TENDON THAT JOINS THE ANIMAL TO HER SHELL. BLIND PRODDING AND POKING IN THE HOLE WILL YIELD LITTLE, HOWEVER. IN-STEAD, RINSE THE CONCH TO CLEAR AWAY THE SHELL CHIPS AND PROBE GENTLY AGAINST THE PINK COLUMELLA (TOWARDS THE CENTER OF THE SHELL), LIFTING THE MEAT WITH YOUR KNIFE. WHEN YOU HAVE DISCOVERED THE JOINT, SEPARATE IT WITH THE KNIFE FROM THE SMOOTH CENTER COLUMN ALONG ITS ENTIRE LENGTH. IT EX-TENDS FROM YOUR NEWLY MADE HOLE STRAIGHT DOWN THE COLUMELLA — A GOOD TWO OR THREE INCHES.

C. TEST YOUR WORK BY PULLING GENTLY ON THE CONCH'S OPERCULUM- IF SHE DOESN'T COME OUT EASILY, SOME OF THE TENDON IS STILL ATTACHED.

To remove the meat and still keep the shell in its entirety, either for "pretty" or for use as a horn, <u>drill</u> a hole in the spire between the second and third row of nodes—directly up from the siphonal canal opening. Make it large enough to admit an ice pick, with which the muscle attachment is severed. This small hole may easily be plugged with a bit of putty or plaster of Paris.

D. When the animal comes free of the shell, continue to hold it by the operculum attached to the foot. This is your handle throughout cleaning. Don't cut it off until the conch is ready for the pot or grinder. Separate the edible part of the orange mantle from the guts and cut the latter away ("slopping"). There's a strip of mantle left hanging from the large meaty foot. Don't forget to feel through this strip carefully if you're hunting pearl. It may be so well enclosed it'll escape your attention.

E. A scooping cut removes the proboscis and the eye stalks, just like gouging out the eye of a potato.

EDIBLE MANTLE
↓

F. NOW YOU ARE FACED WITH REMOVING THE TOUGH ELEPHANT-LIKE SKIN, AND HERE IS WHERE ISLAND TECHNIQUES VARY WIDELY. THE DEXTROUS NATIVE, WITH HIS WORK-HARDENED HANDS, MAKES A SINGLE CIRCUMCISION THROUGH THE SKIN JUST BELOW THE OPERCULUM, FOLLOWED BY A SLIT DOWN TO THE SKIRT. USING HIS THUMB AND THE KNIFE BLADE TO GRASP THE SKIN AT THE INTERSECTION OF THE TWO CUTS, HE PEELS IT WITH A SEEMINGLY CARELESS GESTURE – LIKE PULLING OFF A GLOVE.

G. IT'S EASIER IF YOU FOLLOW THE CIRCUMCISION WITH VERTICAL CUTS AT EACH SIDE OF THE "STEAK", THEN PEEL THE SKIN IN TWO HALVES. IF YOU FIND THAT DIFFICULT, EVEN WITH PLIERS, MAKE SEVERAL MORE CUTS PARALLEL TO THE FIRST TWO AND PULL THE SKIN OFF IN EASY NARROW STRIPS.

H. AN ALTERNATE METHOD IS TO SHAVE OR PARE OFF THE SKIN AS THOUGH PEELING A POTATO.

FOR THE SAKE OF SHOWMANSHIP, SOME GUY WILL USE HIS TEETH TO PEEL THE SKIN FROM A CONCH, WHICH ONLY GOES TO SHOW HE HAS STRONG TEETH AND DOESN'T MIND SLIME.

(THE ANIMAL NEWLY OUT OF THE
SHELL IS COVERED WITH THIS
STICKY GELATINOUS STUFF. IT'LL GET
ON YOUR FINGERS, KNIFE, THE DECK,
THE BOTTOM OF THE DINGHY — WHEREVER
YOU HAPPEN TO BE WORKING. LIME, LEMON
JUICE, OR VINEGAR WILL CUT THIS EFFECTIVELY.)
   VOILA! THE WORST IS
OVER AND YOU HAVE A COM-
PACT ALABASTER-WHITE
PIECE OF DELICIOUS
MUSCLE — READY FOR
THE TENDERIZING
HAMMER, GRINDER,
FOOD PROCESSOR,
OR STEW POT. IN ANY
RECIPE CALLING FOR
GROUND OR DICED CONCH,
YOU MAY INCLUDE THE ORANGE
MANTLE WITH THE STEAK. WHEN
MAKING CRACKED CONCH OR
SEVICHE, FEED THE   MANTLE TO
THE CAT.
   IF KEEPING
THE CONCH FOR
FUTURE USE, NOW
IS THE TIME TO
FREEZE IT UP.

UNLESS YOUR RECIPE CALLS FOR GRINDING, CONCH MEAT SHOULD BE POUNDED TO BREAK DOWN THE MUSCLE FIBRE. THE FLESH AT THE OPERCULUM END OF THE FOOT IS THE TOUGHEST, SO YOU'LL NEED TO "BRUISE" IT THE HARDEST. THE BEST POUNDER IS A METAL MEAT TENDERIZER. A WOODEN ONE, A MALLET, OR THE BOTTOM OF A BOTTLE WILL DO THE JOB, AS WILL A STONE, A BOARD, OR A CLUB. THE MAIN THING TO REMEMBER IS THAT YOU MUST POUND PLENTY IF YOU WANT TENDER CONCH.

EXPERTS SAY YOU SHOULD POUND THE MUSCLE OUT TO TWICE ITS ORIGINAL SIZE. IN DESCRIBING A HARD FIGHT BETWEEN TWO GRENADIANS, A BYSTANDER SAID "HE BEAT HIM LIKE A LAMBIE!"

House Cleaning tips

IF YOU DESIRE THE SHELL
WITHOUT A HOLE IN IT (FOR A COLLEC-
TION OR DECORATION), REMOVE THE
MEAT BY BOILING OR FREEZING.
YOU MAY COOK THE SHELL IN A LARGE
POT, SPIRE DOWN. SIMMER ABOUT
TEN MINUTES AND YOU'LL BE ABLE TO
PULL THE MEAT FREE, USING THE OPERCULUM AS A
HANDLE. OR STAB THE FOOT WITH A HEAVY FORK AND
UNSCREW THE ANIMAL AS HE EMERGES. THIS MEAT
MAY BE USED IN RECIPES CALLING FOR CONCH TO BE
GROUND OR STEWED. ONCE CONCH HAS BEEN BOILED,
IT ISN'T AS EASY TO BREAK DOWN THE MUSCLE
FIBRES BY POUNDING. CHECK TO SEE IF THE ENTIRE COIL
OF GUT IS ATTACHED TO THE END OF THE FOOT. IF NOT,
REPEATED FLUSHINGS AND SHAKINGS ARE THE MOST
PRACTICAL CURE. A SHELL WITH DECAYING CONCH IN-
NARDS IS MOST UNPLEASANT. IF FREEZING IS CHOSEN,
A COUPLE OF HOURS IS NOT ENOUGH. LEAVE HER OVER-
NIGHT, THAW, THEN PULL HARD. IF THIS BREAKS THE
MUSCLE FREE, BUT LEAVES THE GUT INSIDE, PRO-
CEED WITH FLUSHINGS AND SHAKINGS AS
ABOVE. FROZEN CONCH MEAT IS GOOD
FOR ANY CULINARY EFFORT, AND
DEVOTEES OF THIS METHOD
SAY THE FLESH IS MORE
TENDER FOR THE FREEZING.

NOW YOU'VE GOT YOUR CONCH SHELL. THE QUEEN
IS OUT AND YOU WANT TO PUT HER
HOUSE ON THE MANTLEPIECE OR TABLE
FOR SHOW. BUT ITS GOT BITS OF
STUFF ON THE OUTSIDE: WORM
TUBINGS AND CORALINE
GROWTHS, PLUS PATCHES OF
PERIOSTRACUM (THE DARK
SKIN-LIKE OUTER LAYER THAT
PROTECTS THE CALCAREOUS
SHELL FROM BEING DISSOLVED
BY CARBONIC ACID IN SEA
WATER). WHAT TO DO? YOU CAN BRUSH IT AND SCRUB
IT AND PICK AT IT WITH A PENKNIFE. YOU CAN SOAK IT IN
A TEN PER CENT SOLUTION OF BLEACH WHICH WILL
SOFTEN THE ENCRUSTATIONS   AND BRIGHTEN
                                    THE SHELL.

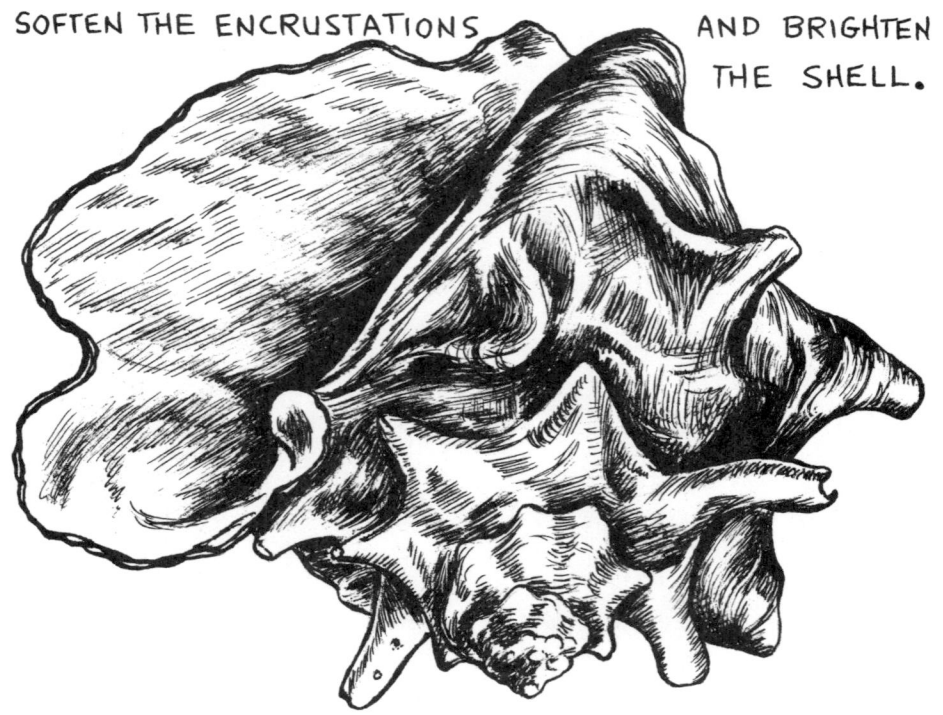

NO, BLEACH WON'T AFFECT THE PINK HUE. (SUNLIGHT FADES THE QUEEN'S COLORS MORE QUICKLY THAN ANYTHING ELSE. IN TIME, INDIRECT LIGHT WILL ALSO CAUSE FADING). MURIATIC ACID IS USED BY MANY SHELL COLLECTORS AND IT IS PROBABLY THE MOST EFFECTIVE WAY TO QUICKLY CLEAN THE QUEEN'S HOUSE. YOU MUST BE VERY CAREFUL, HOWEVER, AS THE ACID WILL EAT INTO THE SHELL ITSELF IF LEFT ON OVER LONG. WEARING EYE PROTECTION, DIP AN OLD TOOTHBRUSH INTO A SMALL AMOUNT OF MURIATIC AND BRUSH IT ON THE OFFENDING AREAS. THE ACID WILL FOAM UP. RINSE IT IMMEDIATELY. WHEN YOUR SHELL IS CLEANED TO YOUR SATISFACTION, WAX OR OIL IT PERIODICALLY TO BRING OUT THE COLOR AND SHINE.

The Queen's Horn

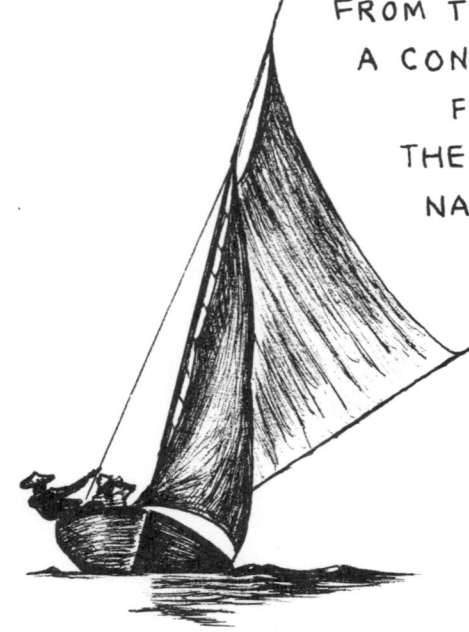

Since neolithic times, blowing a conch horn has been an important form of communication. Its uses were many, from deeply religious to simple and practical. They were used as battle horns. A blast might proclaim a death or warn of danger. Signals were made from village to village. In 1539 the Indians of Florida met De Soto's troops beating drums and blowing conch shells. West Indian slaves were called from the cane fields by a conch horn.

Fishermen still use the conch horn to signal each other across the water. In the Carenage at Grenada, the conch horn announces fish for sale.

CLAY IMAGE FROM COLIMA, MEXICO

YACHTSMEN SOMETIMES
KEEP A CONCH HORN ON
BOARD AS A CONVERSA-
TION PIECE, AND USE
IT IN FOG OR TO BLOW
FOR BRIDGE OPENINGS.
    TOURISTS ARE OFTEN
SEEN WITH PUFFED RED CHEEKS,
FUTILELY TRYING TO BRING FORTH ANY SOUND AT ALL.
    TO MAKE YOUR OWN HORN, CUT OFF THE TOP
INCH OF THE SPIRE WITH A HACKSAW. IN SO
                    DOING,   YOU EXPOSE
                                          THE COL-
                                          UMELLA.

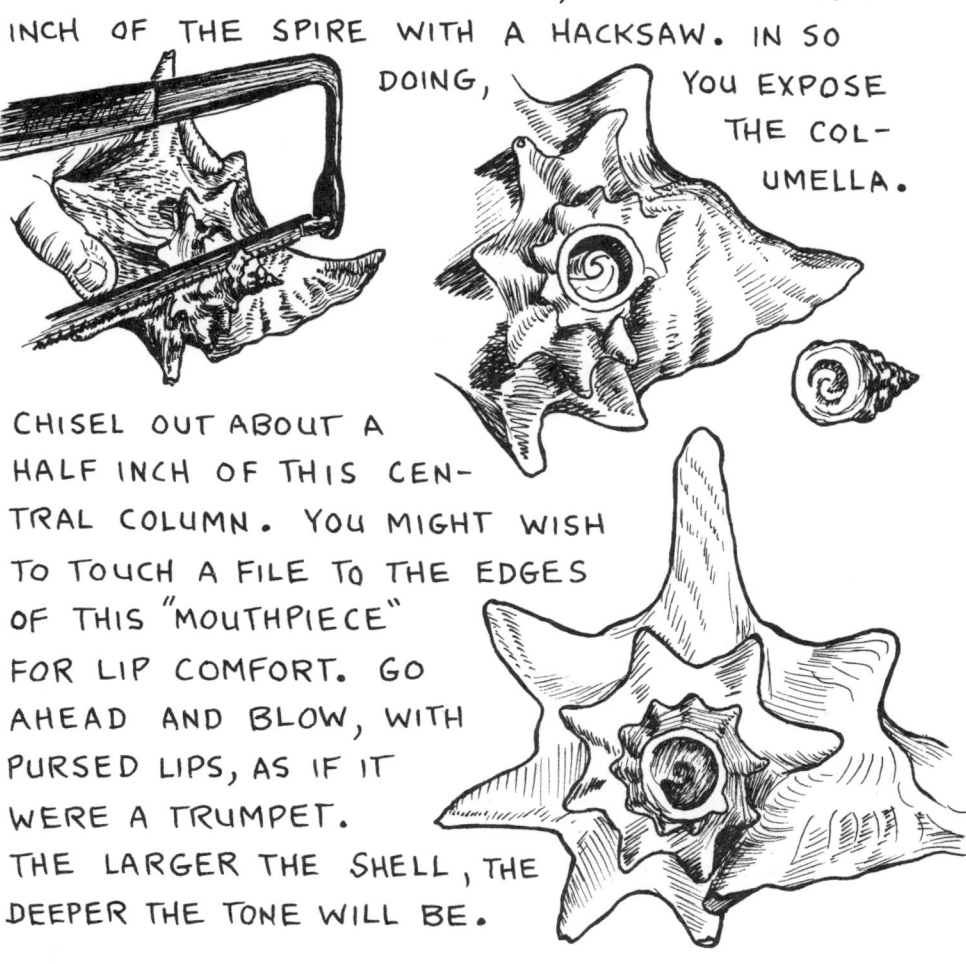

CHISEL OUT ABOUT A
HALF INCH OF THIS CEN-
TRAL COLUMN. YOU MIGHT WISH
TO TOUCH A FILE TO THE EDGES
OF THIS "MOUTHPIECE"
FOR LIP COMFORT. GO
AHEAD AND BLOW, WITH
PURSED LIPS, AS IF IT
WERE A TRUMPET.
THE LARGER THE SHELL, THE
DEEPER THE TONE WILL BE.

The Queen's Table

IF HER HANDSOME SHAPE AND REGAL SIZE WERE NOT ENOUGH, THE BRILLIANT SHADES OF THE SUN WITH WHICH SHE PAINTS HER LIPS — GOLD, FLAME AND ROSE RED — WOULD SURELY PROCLAIM HER FIT TO BE ROYALTY. NOR IS HER BEAUTY ONLY SKIN-DEEP. THERE'S THAT WONDERFUL MEAT WITH ITS INDESCRIBABLE FLAVOR.

CONCH IS ONE OF THE TASTIEST SEAFOODS IN THE WORLD (WHEN PROPERLY PREPARED). IT'S BEEN LAUDED FOR ITS ELEGANCE AND VERSA-TILITY BY GOURMETS THE WORLD OVER, FOR RAW OR COOKED IT'S EQUALLY DELICIOUS. ON THE OTHER HAND, CONCH HAS BEEN MALIGNED AND COMPARED IN TEXTURE TO OLD RUBBER (IMPROPERLY PREPARED). TO THE PEOPLE WHO DINE ON CONCH REGULARLY, IT'S NOT ONLY A DIETARY STAPLE, BUT IS BELIEVED TO HAVE DEFINITE TONIC PROPERTIES. MAYBE PONCE DE LEON LOOKED IN THE WRONG PLACE FOR THE FOUNTAIN OF YOUTH!

WHAT DOES CONCH TASTE LIKE?
IT HAS FIRM, SWEET, CHEWY
FLESH THAT IS MORE REMINISCENT
OF ABALONE THAN ANYTHING ELSE .... OR
MAYBE SCALLOPS, OR CLAMS? YOU MUST
TRY IT YOURSELF TO DECIDE.

EVERY ISLAND AND EVERY FAMILY
HAS ITS OWN SPECIAL AND FAVORITE WAYS
TO BRING THE QUEEN TO THE TABLE. TRADI-
TIONAL RECIPES ARE GIVEN HERE AS WELL AS
SOME THAT ARE NOT SO TRADITIONAL. FOR
YOUR EATING PLEASURE, WE PRESENT ---

THE QUEEN'S TABLE

SCORCHED CONCH

1 CONCH PER PERSON   SALT
SOUR ORANGE OR LIME JUICE
HOT RED PEPPERS OR TABASCO SAUCE

USING A SHARP KNIFE, SCORE THE SURFACE OF THE FRESH RAW CONCH IN A CRISSCROSS PATTERN. SPRINKLE WITH SOUR ORANGE OR LIME JUICE AND RUB WITH RED PEPPERS OR TABASCO. SALT TO TASTE.

CONCH SEVICHE   (SERVES 6 – AS SNACKS)

2 CONCH, POUNDED WELL AND CUT THINLY ACROSS THE GRAIN (PARALLEL TO THE OPERCULUM) IN $1/16$-INCH SLICES.

½ ONION, THINLY SLIVERED.

COVER THE ABOVE WITH A MIXTURE OF ONE HALF WATER AND ONE HALF LIME JUICE, TO WHICH HAS BEEN ADDED: SALT AND PEPPER TO TASTE AND A DASH OF TABASCO. ALLOW TO STEEP AT LEAST ONE HALF HOUR, EITHER IN THE REFRIGERATOR OR AT ROOM TEMPERATURE. DRAIN TO SERVE (WITH CRACKERS). FISH CONCH PIECES OUT WITH TOOTHPICS.

BAGGED CONCH

BAHAMIANS OFTEN CARRY A BAGGIE WITH A KEY LIME, BIRD PEPPER, AND PERHAPS A TOMATO, TO THE FISHING GROUNDS. THE FIRST CONCH CAUGHT IS CLEANED, DICED, AND SEALED UP IN THIS MARINADE – TO BE OPENED LATER FOR A FRESH CONCH SALAD.

CONCH SALAD #1 (SERVES 6)

3 OR 4 CONCH (1 LB.), CLEANED, BRUISED, AND CUT IN SMALL PIECES, SLIVERED, OR SHREDDED.

- 1 MEDIUM ONION, CHOPPED
- ½ C CHOPPED CELERY, INCLUDING A FEW LEAVES
- ½ SMALL GREEN PEPPER, CHOPPED
- 1 LARGE TOMATO, CHOPPED
- ½ C SOUR ORANGE, LEMON, OR LIME JUICE
- SALT, PEPPER, AND HOT PEPPER TO TASTE

(HOT PEPPER — TABASCO OR CRUSHED BIRD PEPPER) MIX WELL AND CHILL AT LEAST ONE HOUR.

CONCH SALAD #2 (SERVES 6)

3 OR 4 CONCH (1 LB.), CLEANED, BRUISED, AND CUT IN SMALL PIECES, SLIVERED OR SHREDDED.

- 1 MEDIUM ONION, CHOPPED
- ½ C GREEN PEPPER, CHOPPED
- 2 TBLSP. PIMIENTO, CHOPPED
- 1 TSP. PARSLEY, CHOPPED
- 1 CLOVE GARLIC, MINCED
- ½ C SOUR ORANGE OR LIME JUICE
- 1 TBLSP. VINEGAR
- 1 TSP. WORCESTERSHIRE
- ½ C OLIVE OIL
- DASH TABASCO
- 1 TSP. SALT
- ½ TSP. OREGANO
- ¼ TSP. THYME

COMBINE ALL INGREDIENTS IN A BOWL AND CHILL SEVERAL HOURS. THIS MIXTURE MAY BE SERVED ON CRACKERS, LETTUCE LEAVES, OR AS A STUFFING FOR TOMATO OR AVACADO.

SALAD NOTES

IN MAKING EITHER OF THE PRE-
CEDING SALADS, YOU MAY SUBSTITUTE
VEGETABLES OF YOUR CHOICE — DICED YELLOW
SQUASH, CUCUMBERS, OR SCALLIONS TO NAME A FEW.
AND, OF COURSE, THE AMOUNT OF HOT PEPPER OR
TABASCO USED IS ACCORDING TO YOUR OWN TASTE.

LAMBIE (SERVES 6)

3 OR 4 CONCH (1 LB.), CLEANED, BRUISED AND
BOILED IN WATER TO COVER FOR ONE HOUR. (OR
PRESSURE COOKED FOR 20 MINUTES) ½ TSP. SALT
   2 HARD BOILED EGGS, CHOPPED   ⅛ TSP. BLACK
   2 TBLSP. CHOPPED GREEN PEPPER   PEPPER
   1 TBLSP. CHOPPED SWEET PICKLE OR CAPERS (OPTIONAL)
   2 TBLSP. SOUR ORANGE OR LIME JUICE

DRAIN CONCH AND GRIND WITH THE MEDIUM BLADE
OF A MEAT GRINDER. MIX WITH THE REMAINING
INGREDIENTS AND SERVE ON A BED OF LETTUCE
AS A SHORT COURSE.

CONCHTAILS

IF YOU FIND CANNED CONCH IN THE CARIBBEAN,
IT MAY BE SERVED, DRAINED, AT COCKTAIL TIME
WITH TOOTHPICS AND A DIPPING SAUCE. OR YOU
MAY BOIL POUNDED CONCH IN WATER TO COVER
FOR AN HOUR (20 MINUTES BY PRESSURE COOKER),
DRAIN, DICE, AND SERVE IN THE SAME MANNER.

CONCH FRITTERS #1 (SERVES 4)

2 LARGE CONCH, INCLUDING THE ORANGE MANTLE

½ ONION

1 EGG

¾ C FLOUR

½ TSP. THYME

1 TSP. BAKING PWD.

FEW DROPS TABASCO

ANY SMALL AMOUNTS OF SPROUTS OR OTHER VEGETABLES (OPTIONAL)

- - - - - - - - - - - - - - - -

CRANK CONCH THROUGH THE FINE CUTTER OF A MEAT GRINDER, FOLLOWED BY THE ONION. MIX ALL INGREDIENTS AND ADD ENOUGH WATER TO MAKE A BATTER SOMEWHERE BETWEEN THE CONSISTENCY OF A BREAD DOUGH AND A PANCAKE BATTER. DROP BY TSP. INTO HOT OIL (365°). COOK UNTIL BROWN. TYPICALLY SERVED AS AN HORS D'OEUVRE WITH A SEAFOOD SAUCE.

CONCH FRITTERS #2 (SERVES 4)

2 LARGE CONCH, INCLUDING THE ORANGE MANTLE

½ GREEN PEPPER

1 SMALL ONION

1 STALK CELERY

½ C MILK

2 TBLSP. TOMATO PASTE

1 TSP. BAKING PWD.

1 C FLOUR

½ TSP. SALT

DASH TABASCO

1 EGG, SEPARATED

PUT CONCH, SWEET PEPPER, ONION, AND CELERY THROUGH THE FINE CUTTER OF A MEAT GRINDER. ADD THE REMAINING INGREDIENTS, EXCEPT FOR THE EGG WHITE. BEAT THE LATTER UNTIL STIFF AND FOLD INTO BATTER. DEEP FRY AS IN FRITTERS #1.

Sauces and Dips

OLD SOUR

(A TRADITIONAL SUBSTITUTE FOR FRESH CITRUS IN CONCH RECIPES)

1 C LIME, SOUR ORANGE OR LEMON JUICE

1½ TSP. SALT   1 BIRD PEPPER, 1 CHILI PEPPER
                    OR CAYENNE TO TASTE

COMBINE INGREDIENTS IN A CLEAN BOTTLE WITH A TIGHT LID. SHAKE WELL AND LET STAND AT ROOM TEMPERATURE FOR 5-10 DAYS. THIS MIXTURE WILL KEEP INDEFINITELY ON THE SHELF. "OLD SOUR" IS PUT UP DURING CITRUS SEASON TO CARRY A FAMILY THROUGH UNTIL THE FOLLOWING YEAR. IT'S USED ON CONCH AND IN DRESSINGS.

RED SAUCE

½ C CATSUP   1 TBLSP. WORCESTERSHIRE

¼ TSP. SALT   1 TBLSP. LIME JUICE

⅛ TSP. PEPPER   DASH TABASCO

MIX ALL INGREDIENTS. MAKES ⅔ C.

TARTARE SAUCE

⅔ C MAYONNAISE   1 TSP. PICKLE RELISH

1 TSP. DIJON MUSTARD   3 CHOPPED STUFFED OLIVES

1 TSP. MINCED ONION   1 FINE CHOPPED HARD

1 TSP. MINCED PARSLEY   BOILED EGG

1 TSP. CHOPPED CAPERS

MIX ALL INGREDIENTS. THIN WITH LEMON JUICE IF NECESSARY.   MAKES 1 C.

WHITE CONCH CHOWDER

2 LARGE CONCH, INCLUDING
   THE ORANGE MANTLE
2 SLICES BACON, CHOPPED
2 ONIONS, CHOPPED   2 TBLSP. LIME JUICE
1 CLOVE GARLIC, MINCED   SALT & PEPPER
2 POTATOES, CHOPPED   2 TSP. PARSLEY
2 CARROTS, SLICED   2 TBLSP. WHITE WINE
3 C WATER   1 CAN (6 OZ.)
3 CHICKEN BOUILLON CUBES   EVAPORATED MILK

CRANK CONCH THROUGH THE MEDIUM CUTTER OF A
MEAT GRINDER. STIR IN LIME JUICE AND 1/2 TSP. SALT.
SET ASIDE TO MARINATE. SAUTÉ BACON AND ADD
ONIONS, GARLIC, POTATOES, AND CARROTS. ADD THE
CONCH AND STIR-FRY A FEW MINUTES. ADD WATER
AND BOUILLON. SIMMER ONE HOUR (OR PRESSURE
COOK 15 MINUTES). ADD SALT AND PEPPER TO
TASTE, PARSLEY AND WINE (IF DESIRED). ADD
MILK BEFORE SERVING.   (SERVES 4.)

ONE, TWO, THREE CONCH CHOWDER (SERVES 4)

IF YOU HAPPEN TO COME ACROSS CANNED CONCH
(COMMONLY FOUND IN PUERTO RICO AND THE VIRGIN
ISLANDS) HERE'S A QUICK AND EASY CHOWDER:
1 CAN (10 1/2 OZ.) CONCH, DRAINED
1 CAN (1 LB.) GERMAN POTATO SALAD
1 CAN (1 LB.) STEWED TOMATOES
MIX TOGETHER AND HEAT. IF IT SEEMS TOO THICK,
THIN WITH CHICKEN BROTH.

RED CONCH CHOWDER (SERVES 4)

2 LARGE CONCH, INCLUDING THE
   ORANGE MANTLE
2 TBLSP. LIME JUICE
   SALT AND PEPPER   DASH TABASCO
2 STRIPS BACON
2 STALKS CELERY, CHOPPED   ½ C WATER
  (INCLUDING LEAVES)   ½ BAY LEAF
1 ONION, CHOPPED   ½ TSP. THYME
1 GREEN PEPPER, CHOPPED   SHERRY
1 (1 LB.) CAN STEWED TOMATOES
½ CAN (6 OZ.) TOMATO PASTE

PUT CONCH THROUGH THE MEDIUM CUTTER OF A MEAT
GRINDER. STIR IN LIME JUICE, ½ TSP. SALT, ⅛ TSP.
PEPPER, AND A DASH OF TABASCO. SET ASIDE TO
MARINATE. SAUTÉ BACON AND ADD CELERY, ONION, AND
GREEN PEPPER. ADD THE CONCH AND STIR-FRY A
FEW MINUTES. ADD TOMATOES, TOMATO PASTE, WATER,
BAY, AND THYME. SIMMER ONE HOUR (OR PRESSURE
COOK 15 MINUTES). REMOVE BAY AND ADJUST
SEASONING. ADD A TSP. OF SHERRY TO INDIVIDUAL
BOWLS IF DESIRED.

NOTE: IN THESE RECIPES, CONCH MAY BE CUBED
OR DICED INSTEAD OF GROUND. IF YOU PREFER
THE FORMER, BRUISE THE MEAT BEFORE CUTTING.
THEN PROCEED WITH THE RECIPE. IT WILL ALSO
REQUIRE A SLIGHTLY LONGER COOKING TIME.

This is a traditional Bahamian dish, used for either breakfast or dinner, depending on the side dishes. It's served with: grits, toast, noodles, rice, spaghetti, or potatoes.

STEAMED (OR STEWED) CONCH (SERVES 4)

4 CONCH, CLEANED AND POUNDED

BOILING WATER

1 TSP. VINEGAR

2 TBLSP. BUTTER

2 TBLSP. OIL

1 ONION, CHOPPED

1 GREEN PEPPER, CHOPPED

1/4 C HAM, CHOPPED FINE (OPTIONAL)

2 C TOMATOES, CHOPPED

1 HOT PEPPER, CHOPPED FINE

1/4 C CATSUP

1 TBLSP. WORCESTERSHIRE

2 C WATER

1/2 TSP. THYME

1 TSP. SALT

1/2 TSP. BLACK PEPPER

DASH LIME JUICE

PLACE CONCH IN BOILING WATER. WHEN WATER FOAMS UP, DRAIN AND ADD FRESH. BOIL AND REPEAT UNTIL THE WATER IS CLEAR. ADD VINEGAR AND SIMMER 20 MINUTES OR UNTIL THE CONCH IS FORK-TENDER. DRAIN AND CUT CONCH INTO BITE SIZE PIECES. IN A HEAVY PAN, HEAT OIL AND BUTTER. SAUTÉ ONION, GREEN PEPPER, CONCH, AND HAM. WHEN THE ONION IS SOFT, ADD TOMATOES, HOT PEPPER, CATSUP, WORCESTERSHIRE, WATER, THYME, SALT, AND BLACK PEPPER. COVER AND STEAM 10-15 MINUTES. STIR IN LIME JUICE.

CRACKED CONCH (SERVES 4)

4 CONCH, CLEANED AND WELL POUNDED
1/4 C LIME, SOUR ORANGE, OR LEMON JUICE
DASH TABASCO   1 TBLSP. WATER
SALT AND PEPPER TO TASTE   1 C CRACKER
1 EGG, SLIGHTLY BEATEN _ _ _ _ MEAL *

MARINATE CONCH IN LIME JUICE AND TABASCO FOR THIRTY MINUTES. DRAIN. DIP CONCH IN THE COMBINED EGG AND WATER, THEN ROLL IN SEASONED CRACKER MEAL. COVER THE BOTTOM OF A HEAVY SKILLET WITH OIL AND FRY CONCH ON EACH SIDE UNTIL GOLDEN BROWN. SERVE WITH LIME, HOT SAUCE AND CHIPS.

*SUBSTITUTIONS FOR CRACKER MEAL: 1 C CORN FLAKE CRUMBS, 1 C FINE BREAD CRUMBS, OR 1/2 C FLOUR.

NOTE: SOME COOKS PARBOIL THEIR CONCH FOR 15-20 MINUTES (OR PRESSURE COOK IT) BEFORE FRYING — TO BE SURE IT WILL BE TENDER. THIS SHOULDN'T BE NECESSARY IF THE CONCH IS PROPERLY PREPARED, AND WHEN PRECOOKED, THE FLAVOR ISN'T QUITE THE SAME.

ON "BRUISING" CONCH INDOORS: DON'T SPLATTER THE KITCHEN OR GALLEY WITH CONCH — PLACE CANVAS OVER THE MEAT FIRST — AND THEN POUND AWAY WITH GUSTO!

BREADED CONCHLETTES (SERVES 2)

2 CONCH, CLEANED AND WELL POUNDED
1 TBLSP. LIME JUICE   1/4 C FLOUR
SALT AND PEPPER   2/3 C ITALIAN-STYLE
DASH TABASCO   BREAD CRUMBS
1 EGG, SLIGHTLY BEATEN   2 TBLSP. OLIVE OIL
1 TBLSP. WATER _ _ _ _ LIME WEDGES

SPRINKLE CONCH WITH LIME JUICE, SALT, PEPPER, AND TABASCO. MARINATE THIRTY MINUTES. PAT CONCH DRY AND CUT INTO 1 1/2-INCH PIECES. DREDGE IN FLOUR, DIP IN COMBINED EGG AND WATER, AND ROLL IN ITALIAN CRUMBS. LET STAND FOR TWENTY MINUTES TO BOND COATING. SAUTÉ QUICKLY IN HOT OLIVE OIL UNTIL BOTH SIDES ARE GOLDEN BROWN. SERVE WITH LIME WEDGES AND/OR A SAUCE OF YOUR CHOOSING.

CONCHBURGERS (SERVES 4)

2 LARGE CONCH, CLEANED (INCLUDE ORANGE MANTLE)
1 MEDIUM ONION   1/4 TSP. PEPPER
2 EGGS, BEATEN   1 TSP. PARSLEY
1/2 C RITZ CRACKER CRUMBS   3 TBLSP. BACON
1 TSP. LEMON OR LIME JUICE _ _ DRIPPINGS

PUT CONCH AND ONION THROUGH THE FINE CUTTER OF THE MEAT GRINDER. ADD EGGS, CRUMBS, LEMON OR LIME JUICE, PEPPER, AND PARSLEY. DROP BY TBLSP. INTO HOT DRIPPINGS IN SKILLET. BROWN ON BOTH SIDES. SERVE WITH TARTARE SAUCE.

TWICE-BATTERED CONCH (SERVES 4)

4 CONCH, CLEANED AND POUNDED
    UNTIL LACY
1/4 C LIME JUICE
DASH TABASCO
1/3 C FLOUR

3 TBLSP. CORNSTARCH	1/4 TSP. CURRY PWD.
PINCH BAKING PWD.	2 TBLSP. GRATED COCONUT
1/4 TSP. SUGAR	1 EGG, LIGHTLY BEATEN
1/3 TSP. SALT	ABOUT 1/4 C ICE WATER

MARINATE CONCH THIRTY MINUTES IN LIME JUICE AND TABASCO. DRAIN AND CUT INTO 1 1/2-INCH PIECES. MIX TOGETHER FLOUR, CORNSTARCH, BAKING PWD., SUGAR, SALT, CURRY, AND COCONUT. STIR IN EGG AND ENOUGH ICE WATER TO MAKE A LOOSE BATTER. IMMEDIATELY ADD CONCH AND STIR TO COAT. DROP INTO HOT OIL (365°). COOK UNTIL GOLDEN BROWN. SERVE WITH A RED OR SWEET AND SOUR SAUCE.

CONCH CRISP (SERVES 4)

2 LARGE CONCH, CLEANED	1 EGG
4 MEDIUM POTATOES	2 TBLSP. FLOUR
1 GREEN PEPPER	SALT AND PEPPER
2 ONIONS	TO TASTE

PUT CONCH AND VEGETABLES THROUGH THE COARSE CUTTER OF THE MEAT GRINDER. ADD EGG, FLOUR, AND SEASONINGS. COOK IN BACON FAT IN AN IRON SKILLET UNTIL BROWN AND CRISP ON BOTH SIDES.

SWEET AND SOUR CONCH

3 OR 4 CONCH, CLEANED,
  BRUISED, AND COOKED
2 TBLSP. BUTTER
3/4 C GREEN PEPPER, CHOPPED
1 CAN (16 OZ.) PINEAPPLE CHUNKS
2 TBLSP. BROWN SUGAR
1 TBLSP. CORNSTARCH   1/4 C VINEGAR
1/8 TSP. GINGER   1 TSP. SOY SAUCE
1/2 C PINEAPPLE JUICE _ _ _ _ _ RICE _ _

CUT CONCH INTO BITE-SIZED PIECES. SAUTÉ IN
BUTTER UNTIL LIGHTLY BROWNED. ADD GREEN PEPPER
AND DRAINED PINEAPPLE AND SIMMER FIVE MINUTES.
MEANWHILE, COMBINE BROWN SUGAR, CORNSTARCH,
AND GINGER AND BLEND WITH PINEAPPLE JUICE,
VINEGAR, AND SOY. STIR THIS MIXTURE INTO THE
CONCH AND COOK, WHILE STIRRING, UNTIL THE SAUCE
IS CLEAR AND THICKENED. SERVE OVER PLAIN RICE
WITH ADDITIONAL SOY ON THE SIDE.   (SERVES 4)

CONCHIGLIE SAUCE   (SERVES 4)

2 CONCH, GROUND   1/4 C OLIVE OIL
1 ONION, CHOPPED   SALT, PEPPER TO TASTE
2 CLOVES GARLIC, MINCED   1/2 C CLAM JUICE
1 TSP. OREGANO _ _ _ _ _ _ _ 1/3 C WHITE WINE

SAUTÉ CONCH, ONION, GARLIC, AND OREGANO IN
OIL. ADD SEASONINGS, JUICE, AND WINE. HEAT.
SERVE OVER CONCHIGLIE (SEASHELL PASTA) OR
  SPAGHETTI. TOP WITH PARMESAN.

BIRD PEPPER

CRUSH BIRD PEPPER IN LIME JUICE — DIP CONCH BEFORE COOKING — OR SPRINKLE THIS MIXTURE ON CONCH AFTER COOKING. (CRACKED CONCH, CONCHBURGERS, ETC.)....

A LITTLE BIRD PEPPER GOES A LONG WAY!

BERMUDA CONCH STEW   (SERVES 4)

3 MEDIUM CONCH, CLEANED, POUNDED AND CUT UP
1/4 LB. SALT PORK OR BACON, CHOPPED

3 LARGE BERMUDA ONIONS	2 TBLSP. FLOUR
1 C WATER	2 TBLSP. CATSUP
2 LARGE CARROTS	1 TBLSP. WORCESTERSHIRE
2 LARGE POTATOES	1 TSP. VINEGAR
SPRIG FRESH THYME	SALT, PEPPER TO TASTE
SPRIG PARSLEY	1/4 C DARK RUM OR SHERRY

SAUTÉ PORK OR BACON IN PRESSURE COOKER WITH 1 CHOPPED ONION. WHEN GOLDEN, ADD CONCH AND WATER. PRESSURE COOK FOR THIRTY MINUTES. ADD 2 REMAINING ONIONS, QUARTERED, CARROTS AND POTATOES, CUT IN CHUNKS, THYME AND PARSLEY. COVER WITH WATER AND PRESSURE COOK AN ADDITIONAL FIVE MINUTES. THICKEN GRAVY WITH FLOUR. REMOVE THYME AND PARSLEY. ADD THE REMAINING INGREDIENTS. SERVE HOT WITH PLAIN RICE.

BIRD PEPPER AND LIME
   CONCH AND THYME — MIX
TOGETHER FOR A TASTE SUBLIME.

CONCH CURRY   (FOR 4)

1 ONION, CHOPPED

1 CLOVE GARLIC, MINCED

2 TBLSP. BACON DRIPPINGS

½ TSP. CURRY PWD. OR TO TASTE

1 TBLSP. FLOUR

1 CAN CR. CELERY SOUP, UNDILUTED

1 C SOUR CREAM

1 C COOKED, THINLY
   SLICED CONCH

COOKED RICE

SAUTÉ ONION AND GARLIC IN DRIPPINGS UNTIL
TRANSPARENT. ADD CONCH. BLEND IN CURRY PWD.
AND FLOUR. STIR IN CR. OF CELERY SOUP AND SOUR
CREAM. SIMMER TEN MINUTES. SERVE OVER HOT
PLAIN RICE AND SPRINKLE WITH SEVERAL OR ALL
OF THE FOLLOWING CONDIMENTS:

CRUMBLED BACON

SIEVED HARD BOILED EGG

CHOPPED GREEN PEPPER

CHOPPED PEANUTS

CHOPPED BANANAS

TOASTED COCONUT

MANDARIN ORANGES

RAISINS

FR. FRIED ONION RINGS
   OR
CHOPPED GREEN ONIONS

CHUTNEY

DICED TOMATO

DICED APPLE OR DRAINED

CRUSHED PINEAPPLE

PRESERVED GINGER
   (CHOPPED FINE)

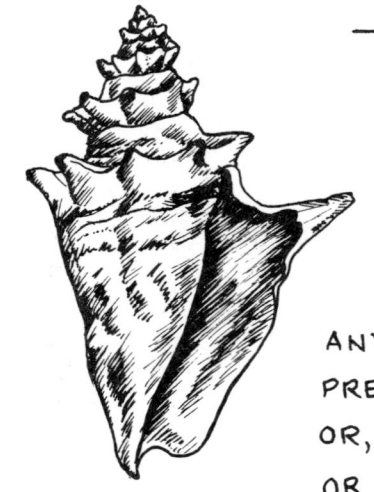

If YOU WANT TO EXPERIMENT WITH CONCH DISHES, TRY ANY SCALLOP RECIPE (AFTER PRESSURE COOKING OR BOILING). OR, SOME COOKS FAVOR CLAM OR ABALONE RECIPES.

THE QUEEN HAS BEEN SPRINKLED WITH MEAT TENDERIZER AND/OR RUBBED WITH PAPAYA LEAF TO SOFTEN THE MEAT. YOU MAY WISH TO TRY —

CONCH QUICHE

CONCH PIZZA

CONCH KEBABS

SCALLOPED CONCH

CONCH *AU GRATIN*

CONCH FOO YOONG

OR   CONCH *A' LA PARISIENNE* —

THE SKY'S THE LIMIT WITH THIS VERSATILE SEAFOOD.

A FEW PAGES FOLLOW FOR NOTING YOUR OWN **CONCH CREATIONS**

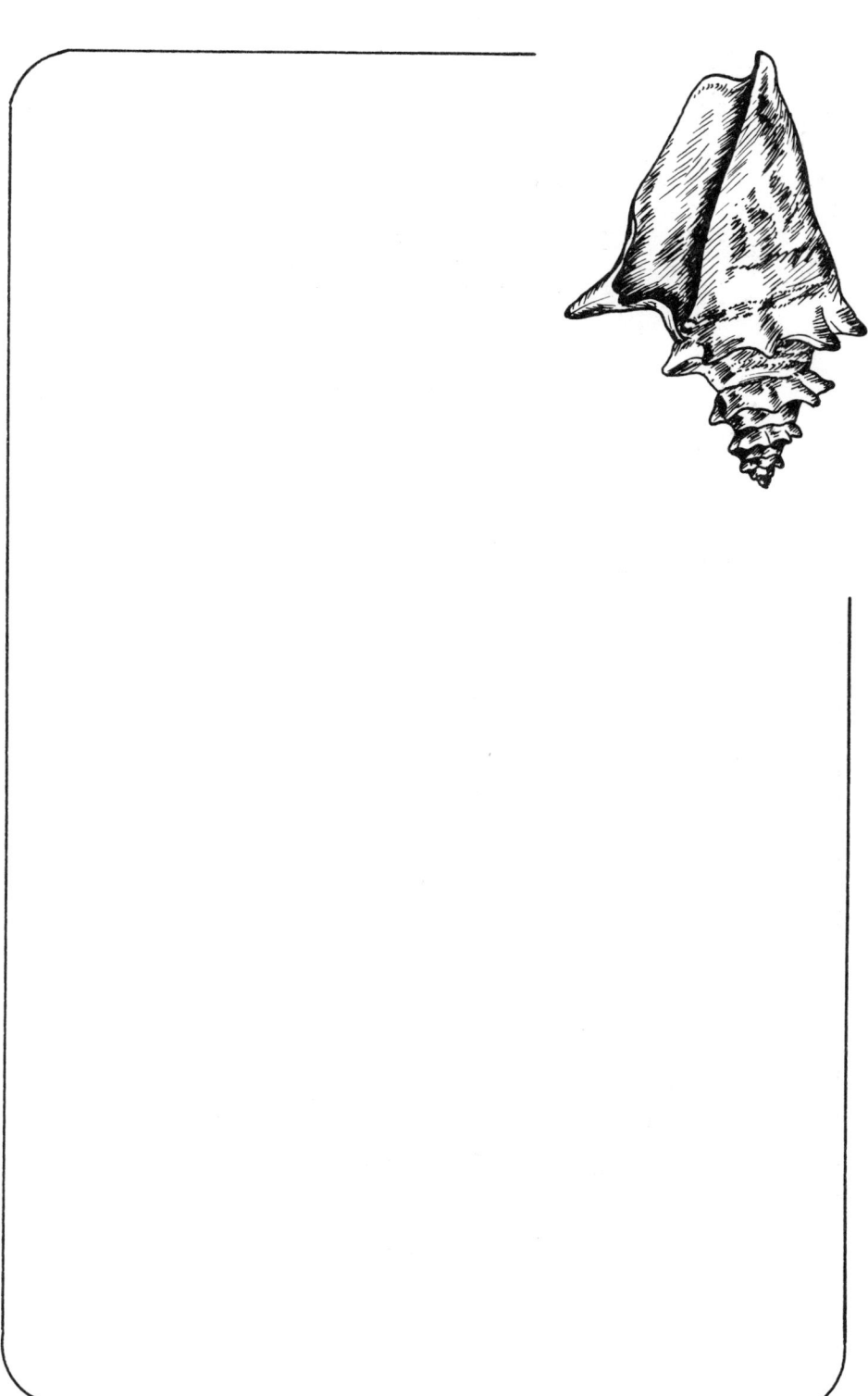

BIBLIOGRAPHY

ABBOTT, R. TUCKER. *SEASHELLS OF NORTH AMERICA.* GOLDEN PRESS, N.Y. 1968.

ABBOTT, R. TUCKER. *SEASHELLS OF THE WORLD.* A GOLDEN NATURE GUIDE. GOLDEN PRESS, N.Y. 1962.

BROWNELL, W.N. AND BERG, C.J. "CONCHS IN THE CARIBBEAN; A SUSTAINABLE RESOURCE?" *SEA FRONTIERS.* INTERNATIONAL OCEANOGRAPHIC FOUNDATION, MIAMI. VOLUME 24, No. 3, 1978.

BROWNELL, W.N. "REPRODUCTION, LAB CULTURE AND GROWTH OF *STROMBUS GIGAS, S. COSTATUS* AND *S. PUGILUS* IN LOS ROQUES, VENEZUELA". *BULLETIN OF MARINE SCIENCE.* 27 (4). UNIVERSITY OF MIAMI. 1977.

BUCHSBAUM, RALPH. *ANIMALS WITHOUT BACKBONES.* UNIVERSITY OF CHICAGO PRESS. 1961.

CAMPBELL, DAVID A. "THE SECRET LIFE OF THE CONCH." *THE EPHEMERAL ISLANDS.* MACMILLAN. LONDON. 1978.

D'ASARO, CHARLES. "ORGANOGENESIS, DEVELOPMENT AND METAMORPHOSIS IN THE QUEEN CONCH, *STROMBUS GIGAS,* WITH NOTES ON BREEDING HABITS." *BULLETIN OF MARINE SCIENCE.* 15 (2). UNIVERSITY OF MIAMI. 1965.

HESSE, C.O. AND KATHY. "CONCH INDUSTRY IN THE TURKS AND CAICOS ISLANDS." *UNDERWATER NATURALIST.* 10 (3).

HIGGS, COLIN. "THE QUEEN CONCH". *BAHAMAS NATURALIST MAGAZINE.* BAHAMAS NATIONAL TRUST. NASSAU.

JOHNSON, WM. R., JR. *BAHAMIAN SAILING CRAFT.* EXPLORATIONS LTD. NASSAU. 1974.

MARSHALL, N.B. *OCEAN LIFE.* MACMILLAN. N.Y. 1971.

MORISON, SAMUEL ELIOT. *ADMIRAL OF THE OCEAN SEA.*
   LITTLE, BROWN AND CO. BOSTON. 1942.

MORRIS, P.A. *FIELD GUIDE TO SHELLS – ATLANTIC & GULF*
   COASTS & W. INDIES. HOUGHTON MIFFLIN. BOSTON. 1947.

POWLES, L.D. *LAND OF THE PINK PEARL.* SAMPSON,
   ETC. LONDON. 1888. SECOND EDITION, MEDIA
   PUBLISHING LTD. NASSAU. 1996.

REA-SALISBURY, VESTA. "COLUMBUS'S ARAWAKS". *SEA*
   FRONTIERS. VOLUME 26, NO. 5. INTERNATIONAL
   OCEANOGRAPHIC FOUNDATION. MIAMI. 1980.

RANDALL, JOHN E. "MONARCH OF THE GRASS FLATS".
   SEA FRONTIERS. VOLUME 9, NO. 3. INTERNATIONAL
   OCEANOGRAPHIC FOUNDATION. MIAMI. 1963.

RANDALL, JOHN E. "THE HABITS OF THE QUEEN CONCH."
   SEA FRONTIERS. VOLUME 10, NO. 4. INTERNATIONAL
   OCEANOGRAPHIC FOUNDATION. MIAMI. 1964.

RAY, CARLETON, AND CIAMPI, ELGIN. *THE UNDERWATER*
   GUIDE TO MARINE LIFE. A.S. BARNES AND CO.
   N.Y. 1956.

SAUL, MARY. *SHELLS, AN ILLUSTRATED GUIDE TO*
   A TIMELESS AND FASCINATING WORLD.
   DOUBLEDAY AND CO. GARDEN CITY, N.Y.

STEVENSON, GEORGE. *KEY GUIDE TO KEY WEST AND*
   THE FLORIDA KEYS. BANYAN BOOKS. MIAMI. 1970.

STIX, HUGH AND MARGARET, AND ABBOTT, R. TUCKER.
   THE SHELL; FIVE HUNDRED MILLION YEARS
   OF INSPIRED DESIGN. HARRY N. ABRAMS, INC.
   N.Y.

BY THE SAME AUTHOR

MAVERICK SEA FARE

A CARIBBEAN COOKBOOK FEATURING TROPI-
CAL FRUITS, VEGETABLES, AND SEAFOOD
SERVED ABOARD A 75-FOOT BRIXHAM TRAWLER.
PEN AND INK DRAWINGS THROUGHOUT.

WINDJAMMER COOKING

WOOD-STOVE COOKERY ABOARD A MAINE
COAST SCHOONER. PEN AND INK SKETCHES
OF SHIPBOARD LIFE, THE GALLEY, AND
SCHOONERS. LARGE RECIPE SECTION.

NARROW WATERS

WATERCOLOR SKETCH LOG OF A CRUISE
DOWN THE INTRACOASTAL WATERWAY. A
BEAUTIFUL GLIMPSE OF COASTAL
CRUISING AT ITS BEST.

AVAILABLE FROM BOOK STORES OR PEN & INK PRESS